NATURAL REMEDIES
Their Origins and Uses

Finn Sandberg
Department of Pharmacognosy
University of Uppsala
Sweden

and

Desmond Corrigan
Department of Pharmacognosy
Trinity College
Ireland

London and New York

First published 2001 by Taylor & Francis
11 New Fetter Lane, London EC4P 4EE

Simultaneously published in the USA and Canada
by Taylor & Francis Inc,
29 West 35th Street, New York, NY 10001

Taylor & Francis is an imprint of the Taylor & Francis Group

Typeset by EXPO, Malaysia
Printed and bound by Gutenberg Press Ltd, Malta

British Library Cataloguing in Publication Data
A catalogue record for this book is available from the British Library

Library of Congress Cataloging in Publication Data
A catalog record for this book has been requested

ISBN: 0-415-27201-7 (hbk)
ISBN: 0-415-27202-5 (pbk)

The photos on the front cover show three phases of the industrial production of natural remedies which can be applied to many plants.

The left photo shows the cultivation of Roseroot, the rhizome of *Sedum roseum* (L.) Scop., syn. *Rhodiola rosea* (L.) in the south of Sweden. The photo in the middle shows the 5 year old rhizome ready to be extracted. The right photo shows the extraction unit in the adjacent factory for preparing the standardized extract for clinical use.

The photos on the front cover were taken by and belong to (left) Ilse Grebe, (middle and right) Swedish Herbal Institute, Gothenburg; the only producer of natural remedies in Sweden having their own cultivation and extraction units.

CONTENTS

Preface

This textbook is the result of a Swedish-Irish collaboration. It is based on a Swedish textbook (published in 1993), which in turn was based on a compendium: Finn Sandberg, *Phytopharmaca therapy*, issued in Uppsala in 1983. Its principle was that the medicinal plants are arranged according to the therapeutic use, *i.e.* the internationally (WHO) accepted ATC-code with some modifications.

The greatest difference from the Swedish edition (1993) is the deletion of much botanical material - at the request of my co-author. The minimum necessary to understand the systematic classification of the drugs is kept whilst reference is made to textbooks in botany.

Personally, I regret the almost non-existing botanical knowledge amongst the Swedish pharmacists of today. Partly, I am responsible for this; during my active period as professor of pharmacognosy (32 years), I have stressed the phytochemical and pharmacological aspects, in contrast to my predecessor, who was a good botanist, and his predecessor, who was Professor Richard Westling, the discoverer of the famous fungus *Penicillium notatum* Westling.

I started the translation and modifications of the Swedish text, writing it by hand. My wife, Ilse Grebe, made it readable with the wordprocessor, and then my co-author gave my text a true English touch. I want to thank my dear wife for extremely good cooperation and typing and my co-author for a brilliant collaboration.

Finally, I want to stress one important fact: the authors are *not* responsible for the stated therapeutic effects of the drugs. We have checked available literature as the basis for our selection of therapeutic medicinal plants.

Finn Sandberg

It has been a privilege and a pleasure for me to have had the opportunity to work with Finn Sandberg on the development of this introductory text on Natural Remedies which form the basis of the rapidly expanding use of phytotherapy and herbal medicine. We hope that it will be used by students in the healthcare professions, particularly by pharmacy students but also by medical students and those enrolled on courses leading to degrees and other qualifications in herbalism and complementary medicine.

The explosion of interest in the use of plant medicines by the general public over the last twenty years has been phenomenal. This, in turn, has encouraged healthcare professionals to take a more active interest in acquiring knowledge about these remedies. Thus a unique science, Pharmacognosy, which was once considered obsolescent by many pharmaceutical scientists and educators, has once again emerged as a vibrant area of academic research, clinical utilisation and public awareness. Medical and pharmaceutical journals, which previously disdained papers and reviews on medicinal plants, have become more accepting of the fact that these medicines do work and that nature is still a remarkable chemist capable of surprising mankind with the diversity and utility of the chemical structures it can produce.

For two pharmacognosists who can recall when it was neither popular nor profitable to be involved with medicinal plants, the sea change in attitude is remarkable and welcome. Indeed, Finn Sandberg is one who must be credited, with other colleagues, with keeping alive the spirit and credibility of research and teaching about plants as drugs and medicines. His is the major contribution to this book through the innovative approach to Chapter VI which presents information on the therapeutic uses of plants using the method developed by WHO as the basis for the chapter. This ATC classification, which uses an Anatomical, Therapeutic and Chemical approach to the cataloguing of synthetic medicinal products, has for the first time been applied (in English) to plant drugs. Originally Finn developed this concept in Swedish. We have taken the opportunity provided by the translation into English to update and expand that original concept and to support the main thrust of Chapter VI through a series of chapters which highlight for students certain aspects of medicinal plants which are important for a full understanding of the role of plant drugs in healthcare.

Chapter III is very much based on the thought expressed by the Spanish-American philosopher Santayana that those who forget their history are condemned to relive it. In describing the issue of quality control of herbal medicinal products we are touching on the original ethos and focus of Pharmacognosy even though the general tone of this book is the clinical usefulness of plants and the phytochemicals derived from them. It is our objective to highlight how quality assurance and control can have a major impact on both the safety and efficacy of medicines from plants. Chapter VIII addresses the issue of the safety of phytomedicines by challenging the potentially dangerous perception that because something is natural it is automatically safe. In describing not only plants which are overtly toxic and harmful, but also those whose consumption carries the risk of dependence or addiction, of carcinogenicity, of hepatotoxicity etc., we argue for an attitude that recognises that plants can be powerfully beneficial agents in illness but can equally be powerfully damaging to humans and animals.

Chapter VII provides an introduction to the new and exciting area of chemoprevention, where plants and their secondary metabolites are increasingly being used to prevent the major killer diseases of the developed world such as cancer and heart disease. Chapter V summarises the philosophies underlying the use of plants in complementary and traditional systems of medicine. We also signpost some of the regulatory issues surrounding the availability of plant derived medicines which are effective, safe and of the highest quality to patients/consumers who choose to use such medicines or who are advised by their pharmacist or other healthcare provider to use them.

The modern world of medicinal plants is a fascinating kaleidoscope of botany, chemistry, medicine, toxicology, anthropology, ethnopharmacology, entrepreneurship, agriculture, information technology and old-fashioned humanity. It is our hope that this introduction to the origins and uses of Natural Remedies, will encourage students to learn more about plants that heal and about how they can be used effectively and safely.

Desmond Corrigan

CHAPTER I
INTRODUCTION

The term Phytotherapy is now accepted in Europe in English, French and German. It means simply a therapy using processed or crude plant material. There are several subdisciplines involved: botany for the identity of the plant used; agricultural knowledge for the cultivation of the drug plant; phytochemistry for the extraction and structure elucidation of active ingredients, pharmacology for the mode of action of active substances; internal medicine for the clinical use of the product.

Phytotherapy is in a way similar to the term "materia medica" of earlier days.

It should be remembered that before the introduction of synthetic chemistry, all remedies were natural products, and the corresponding drugs were included in the various pharmacopoeias. The difference between modern and old phytotherapy is that nowadays extracts of the crude drugs can be standardized to a certain content of active ingredients thus, guaranteeing the correct and reproducible dosage.

Definitions

Materia Medica is an old term from the time when all remedies were natural products of vegetable, animal or mineral origin. According to Schmidt (1759-1809, Vienna) *Materia Medica* included pharmacognosy (description of the drugs and their identification) and pharmacodynamics (the effects of the drugs). Schmidt was the first to use the term pharmacognosy.

The term phytotherapy was first used by the French physician Leclerc and the concept was further developed by Weiss in Germany.

History

The medical use of natural products has been verified among the Sumerians about 4,000 years BC, who used among others Opium. In China, Ginseng, Rhubarb and Ma-Huang (Ephedra) have been used for at least 5,000 years.

The Egyptians had a good knowledge of natural drugs. The "Papyrus of Ebers", dating from about 1550 BC, mentions Aloe, Myrrha, and Henbane.

Hippocrates (460-377 BC), who is called the father of medicine, was the most brilliant of the Greek physicians and he used a great number of drug plants. The Greeks dominated medicine during the Roman period. During the first century AD, Dioscorides described in his work "De Materia Medica" about 500 medicinal plants: their botanical origin, production and use. This book and the works of medic Galenos (second century AD) who used several hundred medicinal plants, dominated until the Middle Ages, when knowledge of medicinal plants was maintained in two ways: by the Arabs and by the Christian monks, who cultivated spices and medicinal plants in the gardens of the monasteries.

Medicinal plants were described in "Kräuterbücher" (Herb books) e.g. by

L. Fuchs, in the 16th century, which were important in the preservation of this valuable knowledge.

In 1803 the German pharmacist Sertürner succeeded in the isolation of Morphine from Opium. This event started a new epoch, focused on isolation and characterisation of pharmacologically active substances from medicinal plants. With the advent of synthetic organic chemistry, interest in the use of medicinal plants declined in Western medicine during much of the 20th century.

In the past decade there has been an increasing public interest in natural products, partly due to a suspicion about the safety of synthetic, non-natural remedies. However, both types should be used, sometimes together, in conventional medicine as well as in complementary or alternative medicine.

Names of drugs used

The therapeutically active ingredients can be obtained from fresh plants, but mainly from dried parts of a plant, called drugs.

The names of the drugs have varied, but in this book we follow the Latin nomenclature of the European Pharmacopoeia, i.e. botanical name in genitive and plant part of the drug in nominative singular, e.g. *Gentianae radix*, Gentian root.

The following terms are used to indicate different parts of a plant:

1. Drugs

Bulbus - bulb, which is a stem surrounded by thick leaves, filled with nutrients and poor in chlorophyll.

Cortex - bark, which is obtained from both stem and root. Bark-drugs are obtained from plants with secondary growth and consist of all tissues outside the cambium.

Flos, Flores - flowers. Consist of separate flowers and/or entire inflorescences.

Folium - leaves. Consist of the middle leaves of the plant.

Fructus - fruit. This name of a drug is not always botanically strictly correct. *Cynosbati fructus*, hip: the main axis of the flower is well developed and is part of the fruit.

Herba - the drug consists of the entire above-ground part of the plant or flowering twigs.

Lignum - wood, which consists of the woody part inside the cambium in plants with secondary growth. The term is not always botanically correct, e.g. in *Quassiae lignum*, where the thin bark is also present.

Pericarpium - fruit-shell. Can also be one part of the fruit shell, e.g. *Aurantiae pericarpium* (bitter orange) consists of the outer layer, called flavedo.

Radix - root. This term is not always botanically correct, e.g. *Gentianae radix* - includes roots and rhizome.

Rhizoma - rhizome, i.e. a subterranean stem, supplied with side roots, e.g. *Valerianae rhizoma*, Valerian root.

Semen - seed, which consists of the seeds isolated from the fruit, e.g. *Strophanthi grati semen*. In some cases the stem of the seed is missing, e.g. *Colae semen*, Cola seeds.

Tuber - a subterranean organ, filled with nutrient. Can be either a root or a subterranean stem. It is thick, consisting of parenchyma with nutrients (generally starch) with small amounts of lignified elements.

2. *Preparations*

Aetheroleum - essential oil, obtained from plant material by watersteam distillation, with a pronounced smell and consisting of a complex mixture of volatile ingredients.

Balsamum - balsam, which is a solution of resin in volatile oil. It is generally formed in special cells in the tree.

Extractum - extract, generally an alcoholic extract, but also other solvents can be used, e.g. ether for *Filicis extractum*. Extracts may be fluid extracts (1 part drug to 1 part extract), soft extracts (75% solids) or dry extracts (95% solids).

Oleum - oil, which is pressed fatty oil from plant material (seed: castor oil or fruits; olive oil) and which does not evaporate at room temperature.

Pyroleum - tar, obtained by dry distillation of plant material.

Resina - which can be obtained from secretory cells from certain plants or from distillation of balsam, when it will be the residue of the distillation.

Tinctura - tincture, which is an alcoholic extract, generally containing the equivalent of 1 part of drug to 10 parts of solvent.

The crude drugs may be obtained from wild or cultivated plants. The tendency has been for some decades to use cultivated plants more as their quality is easier to control. The cost of labour may also be a factor of importance, as can an increased demand which may cause extinction of a wild species.

Cultivated medicinal plants are produced all over the world. In Europe, Germany and Hungary have long traditions in this field.

Factors relating to medicinal plant production which can affect the quality, safety and efficacy of medicinal products based on plants, their preparations or single chemical entities isolated from them are discussed in Chapter III.

Reading List

Kremers and Urdang's History of Pharmacy, Fourth Edition, Revised by Glenn Sonnedecker, Lippencott Company, Philadelphia 1976

J. Berendess: Geschichte der Pharmazie, Leipzig 1898 Proc. 33rd. Intern. Congress, Hist. Pharmacy Stockholm, 1997, p. 43 J.G. Bruhn, Pharmacognosy - "The source and future of pharmacy."

CHAPTER II
BOTANICAL PHARMACOGNOSY

Botany deals with plants, their appearance and function, lineage, geographical distribution, development and history. Botany comprises several parts:

1. MORPHOLOGY describes the appearance of the outer parts of the plants: root, leaves, flowers, fruits and seeds.

2. ANATOMY describes the microscopic structure of different parts of the plant. The study of the anatomical structure is carried out with the aid of light and electron microscopes on thin sections of all parts of the plant. A special branch is the study of pollen of plants, which is called palynology.

3. PLANT PHYSIOLOGY deals with the normal functions of the plants (photosynthesis in the leaves of the green plants and subsequently biosyntheses) and comprises also descriptions of the life cycle of the plants and the external influence of light, humidity, pressure and temperature.

4. SYSTEMATIC BOTANY deals with the classification in units of different ranks and the relationship between these units. These units are called taxa (singular: taxon), which has given the synonymous name taxonomic botany.

The scientific name of the plants is dealt with in the part of systematic botany which is called botanical nomenclature.

An important branch of systematic botany is chemotaxonomy, where the relationship between the plants is based on the occurrence of specific chemical substances.

5. PLANT GEOGRAPHY describes the geographical distribution, which is determined by climate and its changes, history and capability of spreading.

6. PLANT GENETICS comprises the study of the unique genetic constitution of each plant species, its origin and changes, which has application also in medicinal plants.

7. ETHNOBOTANY, which has close connections to pharmacognosy, describes how different cultures have collected or cultivated and used plants as drugs, food, clothes and tools.

For sections 1–3 and 5–7, reference can be made to textbooks of botany. Only section 4 will be dealt with here, which gives the systematics of medicinal plants which might be difficult to find in ordinary textbooks of botany.

Systematic botany

Systematic botany deals with the classification of the plant world in units of different ranks and the relationship between these units.

The classification of plants was for a long time based only on the external morphology, but nowadays both the Theory of Evolution and the results of modern genetic studies have been of decisive importance for the understanding of the processes that lead to changes between and the origin of new taxa.

The need for a system with useful principles of classification has long existed, and the progress of research has changed the character of these systems. Thus, the system, introduced by Linné (Linnaeus), was a purely artificial system for phanerogams, built on the structure of the flower and the number and arrangements of stamens and pistils.

This system was gradually replaced by the one that is now used and which is built on real relationships. Taxa are compared with one another and on these grounds a natural system is made. Two taxa are thus related, if the number of similarities is large. Today a stable plant system of this type is aimed at, but new results of taxonomic research have the consequence that the system is continuously being modified.

The present flora is the result of a long geologic evolution, which is only partly known and will never be fully known, because most ancestors died out without leaving any trace. At best, fossils can give some incomplete information. For an entirely complete picture one is nevertheless left to hypothetical conclusions.

Even if the process of formation of a species is continuous, the species is still the fundamental unit in systematic classification. The border between species is kept by the fact that they do not cross freely with another. Hybrids that may be formed are generally sterile. One species generally consists of a varying number of individuals which are somewhat different from each other, but they can cross freely with another (exchange genes). The smallest functioning biological unit is a population. A population, the size of which can vary, is a group of individuals, among which a change of genes takes place more or less regularly. Each individual gets usually a unique combination of inheritance, shown by its special appearance. Thus, the population is variable. Each individual gets usually a unique combination of genes. For instance, when there are changes in the milieu only the most fitted will survive; this is natural selection. This selection is one of the factors that leads evolution further, whereby new species are created.

The species is difficult to define. Most cases will be covered by the following: plants belong to different species if they have clear differences in morphological character and are effectively isolated from each other.

The nomenclature system, which was introduced by Carl von Linné in 1753 and has been used since then, is the binomial. The name of the species consists of two Latin words, where the first is the name of the genus (commencing with a capital letter) and the other is the epithet of the species; e.g. *Digitalis purpurea*, Purple foxglove. The name of the species cannot be the same as the name of the genus. If the epithet of the species consists of two words, they are connected with a hyphen. The person who first scientifically described a species is called the auctor or the authority for the binomial; thus, *Strychnos nux-vomica* L., where L. means Carl von Linné. Other for abbreviations for auctors are DC for de Candolle, Engl. for Engler, Sol. for Solander, Sparrm. for Sparrman, Thunb. for Thunberg and so on. It is the name of the auctor that makes the Latin name of the species specific or exact.

The rules for nomenclature are ratified by international botanical congresses and are published as *The International Code of Botanical Nomenclature*. The description by the auctor of a species is related to a certain herbarium sheet, which is called the type-specimen. If there is any doubt as to what the author meant with his specimen, it is always possible to check "the type-specimen". For the same reason a phytochemical publication should always contain details of where "a voucher specimen" of the species studied is kept.

The species is the basic unit in systematic botany. Sometimes it is necessary to divide a species into subgroups to describe the variation within the species. The most usual ones are the following, arranged according to diminishing rank:

1. species - abbreviated to sp.
2. subspecies - abbreviated as subsp. or ssp. (geographic race)
3. varietas - abbreviated v. or var. (local race)
4. forma - abbreviated f. (genetically different form that appears in several places).

Taking blueberry as an example, the taxonomic hierarchy with the Latin endings for each category appears as follows:

Category	Latin Ending	Example
Division	- phyta	Spermatophyta
Subdivision	- phytina (= Angiospermae)	Magnoliophytina
Class	- ate (=Dicotyledoneae)	Magnoliatae
Superorder	- anae	Ericanae
Order	- ales	Ericales
Family	- aceae	Ericaceae
Subfamily	- oideae	Vaccinioideae
Tribus (Tribe)	- eae	Vaccinieae
Genus		Vaccinium
Species		myrtillus L.

Survey of the plant kingdom

The plant kingdom contains two main groups of organisms: Prokaryota and Eukaryota.

Prokaryota, which comprise only bacteria and bluegreen algae, have no cell nucleus, which means that the DNA (desoxyribonucleic acid) is not covered by a nuclear membrane.

Eukaryota, which have a cell nucleus, comprise all other organisms from simple algae to highly developed dicotyledons.

Among Eukaryota, it is the Spermatophyta that are of the greatest pharmacognostic interest. Now follows a survey of the classification of the plant kingdom. The numbering is made down to and including superorder.

I. Prokaryota

1. Div. Schizophyta - bacteria
2. Div. Cyanophyta - bluegreen algae

II. Eukaryota

Divisions 3-9 comprise algae.

3. Div. Eugenophyta
4. Div. Pyrrophyta - dinoflagellates
5. Div. Chrysophyta - yellow algae
6. Div. Xanthophyta - yellow-green algae
7. Div. Phaeophyta - brown algae
8. Div. Chlorophyta - green algae
9. Div. Rhodophyta - red algae

10. Div. Mycota - fungi
10.1. Subdiv. Myxomycotina - slime moulds
10.2. Subdiv. Eumycotina - true fungi
10.2.1. Class Ascomycetes - sack fungi
10.2.2. Class Basidiomycetes - basidial fungi

While divisions 1-10 are thallophytes, divisions 11-13 are cormophytes.

11. Div. Bryophyta - mosses
12. Div. Pteridophyta - ferns
12.1. Subdivision: Lycopodiophytina
12.2. Subdivision Equisetophytina
12.3. Subdivision Polypodiophytina
13. Div. Spermatopyta (phaneroganes)
13.1. Subdivision Gymnospermae (Pinophytina)
13.1.2. Class Gnetatae
13.1.3. Class Ginkgoatae
13.2. Subdivision ANGIOSPERMAE (Magnoliophytina)
13.2.1. Class DICOTYLEDONAE (Magnoliatae)
Dicotyledonous Plants, about 165,000 species
13.2.1.1. Superorder Magnolianae
divided into orders and families
Annonales: Myristicaceae
Illiciales: Illiciaceae
Schisandraceae
Laurales: Monimiaceae
Lauraceae
13.2.1.2. Superorder Nyphaeanae
divided into orders and families
Piperales: Piperaceae
13.2.1.3. Superorder Ranunculanae
divided into orders and families
Ranunculales: Menispermaceae
Ranunculaceae
Podophyllaceae
Berberidaceae
Papaverales: Papaveraceae
13.2.1.4. Superorder Caryophyllanae
divided into orders and families
Caryophyllales:
Caryophyllaceae
Phytolaccaceae
Cactaceae
Chenopodiaceae
13.2.1.5. Superorder Polygonanae
divided into orders and families
Polygonales: Polygonaceae
13.2.1.6. Superorder Malvanae
divided into orders and families
Malvales: Sterculiaceae
Filiaceae
Malvaceae
Moraceae
Urticales: Cannabaceae
Urticaceae
Euphorbiales: Euphorbiaceae
Rhamnales: Rhamnaceae
13.2.1.7. Superorder Violanae
divided into orders and families
Violales: Passifloraceae
Violaceae
Caricaceae
Cucurbitales: Cucurbitaceae
Salicales: Salicaceae
Capparales: Brassiacaceae
13.2.1.8. Superorder Theanae
divided into orders and families
Paeoniales: Paeoniaceae
Theales: Theaceae
Hypericaceae
13.2.1.9. Superorder Primulanae
divided into orders and families
Primulales: Primulaceae
Ebenales: Styracaceae
13.2.1.10. Superorders Rosanae
divided into orders and families
Hamamelidales:
Hamamelidaceae
Fagales: Fagaceae
Corylaceae
Betulaceae
Saxifragales: Grossulariaceae
Crassulaceae
Droserales: Droseraceae
Rosales: Rosaceae
Malaceae
Amygdalaceae

13.2.1.11. Superorder Myrtanae
divided into orders and families
Myrtales: Myrtaceae
Onagraceae

13.2.1.12 Superorder Rubanae
divided into orders and familes
Sapindales: Anacardiaceae
Sapindaceae
Fabales: Caesalpinaceae
Mimosaceae
Fabaceae
Rutales: Rutaceae
Simarubaceae
Burseraceae
Meliaceae
Polygalaes: Polygalaceae
Krameriaceae
Geraniales: Geraniaceae
Linales: Linaceae
Erythroxylaceae
Celastrales: Celastraceae

13.2.1.13 Superorder Vitanae
divided into orders and families
Vitales: Vitaceae

13.2.1.14. Superorder Santalanae
divided into orders and families
Santalales: Santalaceae
Viscaceae

13.2.1.15. Superorder Aralianae
divided into orders and families
Araliales: Araliaceae
Apiaceae

13.2.1.16. Superorder Asteranae
divided into orders and families
Asterales: Asteraceae

13.2.1.17 Superorder Solananae
divided into orders and families
Solanales: Solanaceae
Convolvulaceae
Boraginales: Boraginaceae

13.2.1.18. Superorder Ericanae
divided into orders and families
Ericales: Ericaceae

13.2.1.19 Superorder Cornanae
divided into orders and families
Cornales:
Aquifoliaceae
Sambucaceae
Menyanthaceae
Dipsacales: Caprifoliaceae
Valerianaceae

13.2.1.20 Superorder Gentiananae
divided into orders and families
Olales: Oleaceae
Gentianales: Loganiaceae
Rubiaceae
Gentianaceae
Apocynaceae
Asclepiadaceae

13.2.1.21. Superorder Lamianae
divided into orders and families
Lamiales: Scrophulariaceae
Pedaliaceae
Acanthaceae
Plantaginaceae
Verbenaceae
Lamiaceae

13.2.2. Class MONOCOTYLEDONAE (Liliatae) - Monocotyledonous plants, about 55,000 species

13.2.2.1. Superorder Aranae
divided into orders and families
Arales: Araceae

13.2.2.2. Superorder Lilianae
divided into orders and families
Dioscoreales: Dioscoreaceae
Asparagales: Convallariaceae
Agavaceae
Hyacinthaceae
Asphodeliaceae
Alliaceae
Colchicaceae
Lilales: Iridaceae
Melanthiales: Melanthiaceae

13.2.2.3. Superorder Bromelianae
divided into orders and families
Bromeliales: Bromeliaceae

13.2.2.4. Superorder Zingiberanae
divided into orders and families
Zingiberales: Musaceae
Zingiberaceae
Costaceae

13.2.2.5. Superorder Commelianae
divided into orders and families
Arecales: Areaceae

The aforementioned plant families cover the taxonomic distribution of the medicinal plants mentioned in this book. It gives, hopefully, an idea of the taxonomically uneven distribution of medicinal plants. Those readers who are more interested in plant taxonomic aspects are referred to Index Kewensis and similar literature. An understanding of botanical nomenclature and the conventions surrounding it is not an esoteric intellectual pursuit but is integral to the first essential rule relating to medicinal plants, namely their correct identification. This is fundamental to the use of safe, effective and good quality plant-based or derived medicines.

Reading list

Heywood, Vernon H. (ed.) 1993. *Flowering Plants of the World*. 2nd ed. B.T. Batsford Ltd., London. ISBN 0 7134 7422 X.

Cronquist, Arthur 1988. *The Evolution and Classification of Flowering Plants*. 2nd ed. New York Bot. Garden.

Mabberley, D.J. 1997. *The Plant-Book: A Portable Dictionary of the Higher Plants*. 2nd ed. CUP. (Cambridge Univ. Press).

Takhtajan, Armen 1997. *Diversity and Classification of Flowering Plants*. Columbia Univ. Press, 643 pp.

Foster, A.S. & Gifford, E.M. 1959. *Comparative Morphology of Vascular Plants*. 1st ed. W.H. Freeman, San Francisco. 2nd ed. 1974, Freeman, San Fr., 751 pp. 3rd ed. with changed title: Gifford, E.M. & Foster, A. 1989. *Morphology and Evolution of Vascular Plants*, Freeman, etc.

CHAPTER III
THE QUALITY CONTROL OF HERBAL MEDICINAL PRODUCTS

Introduction

In order to use a plant drug as a herbal medicinal product or as a source of a medicine i.e. an isolated purified compound, it is important to ensure that (a) the correct plant is used; (b) that the plant contains the required chemical constituents; (c) that these constituents are present at the required level; and (d) that the type and level of constituents are reproducible from batch to batch of the drug plant.

The very fact that these criteria can be stated suggests that it is possible that plant drugs have failed to meet one or more of the above criteria in the past. One of the major reasons for this state of affairs is the fact that the secondary metabolites responsible for the pharmacological activity of the plant are biosynthesised, involving precursors from photosynthesis and primary metabolism via the shikimic acid, polyketide and mevalonic pathways, in living plants. All living organisms are inherently variable and in plants this variability extends not only to the physical appearance i.e. morphological characters, of the plant but also to the chemical compounds biosynthesised by the plant. This variability affects both the level of secondary metabolite and the overall yield of phytomass (plant bulk), and as a result the quality of the plant. The term "quality" is used to denote a combination of high percentage concentration of desired compound(s) and a high total yield of that compound.

EU and WHO guidelines

Mindful of this variability, the EU has in place two Guidelines concerning the quality of medicinal plants. One relates to the Quality of Herbal Medicinal Products, the other deals with the concept of Good Manufacturing Practice (GMP) as it is applied to the manufacture of those products. The World Health Organisation (WHO) has also published Quality Control Methods for Medicinal Plant Materials and as an integral part of its monographs on selected medicinal plants WHO includes a comprehensive quality monograph on each plant.

In the EU's Quality Guideline the initial control of the herbal drug involves a comprehensive specification for that drug. Ideally that monograph should be from a recognised Pharmacopoeia and increasing numbers of monographs on plant drugs now appear in the European and various national (particularly the US, French, German and Chinese) Pharmacopoeias. Other Pharmacopoeias such as the British Herbal Pharmacopoeia and various Homoeopathic (e.g. British and German) Pharmacopoeias are unofficial but contain valuable quality monographs. These monographs usually include macroscopical and microscopical descriptions of the plant. Identification based on Thin - Layer Chromatography (TLC) and chemical assays are standard items in such monographs.

Where no pharmacopoeial monograph exists, manufacturers must supply their own comprehensive specification which should include the following:

1. Latin botanical name and authority (i.e. L. for Linnaeus, DC for De Candolle etc) and common (i.e. English) name.
2. Information on the site of collection, time of harvest, stage of growth, treatment during growth, drying and storage conditions.
3. Tests for microbiological quality.
4. Tests for residues of pesticides, fumigants, radionuclides, heavy metals, other contaminants and likely adulterants.
5. Assays for content of constituents of known therapeutic activity; where these are not yet known with certainty, assays of marker substances must be used. All of these methods must be fully validated, using standard and accepted protocols for analytical method validation.

In the GMP guideline these items are spelt out in more detail in the section dealing with documentation of the specifications for starting materials. For example in the section dealing with microbial quality the GMP guideline mentions "the tests to determine fungal and/or microbial contamination, including aflatoxins and pest infestations and limits accepted." It further specifies that treatments to reduce such contamination should be documented. Under "Quality Control" the guideline suggests that personnel should have particular expertise in herbal medicinal products in order to carry out identification tests and recognise adulteration, the presence of fungal growth, infestations, non-uniformity within a delivery of crude plants etc.

This chapter attempts to explain why all these requirements are considered necessary by looking at the factors affecting herbal drug quality and how the quality specifications are influenced by these factors.

Factors influencing quality

The factors which influence the quality of medicinal plants are:

1. Identity of the plant
2. Genetic factors
3. Ecological factors
4. Post harvest handling
5. Method of storage
6. Environmental contamination

Identity of the plant

It is obvious that consistent quality for products of herbal origin can only be assured if there is a rigorous and detailed definition of the starting materials. Failure to ensure exact botanical identification of the plant material puts the patient at risk not only of ineffective therapy but also of exposure to potentially toxic plants.

Botanical identification involves both macroscopical and microscopical inspection of the raw material. According to WHO, the macroscopical identity of plant materials is based on the shape, size, colour, surface characteristics, texture, fracture and appearance of the cut surface of leaves, herbs, seeds, fruits, barks, stolons, rhizomes and roots. Literature reports of poisonings and deaths resulting from the consumption of Digitalis collected by mistake for Comfrey (*Symphytum*) and of Oleander instead of Eucalyptus leaf tea highlights the importance of this simple and basic part of the quality procedure.

Clearly macroscopical evaluation on its own cannot guarantee identification or quality and microscopical examination is indispensable, particularly where products have been chopped or powdered. Microscopic identification is based on the presence or absence of certain key tissues, cells and cell inclusions (e.g. xylem, trichomes, stomata, fibres, starch grains, calcium oxalate crystals) which are diagnostic for a particular plant. While many herbal drugs cannot be unequivocally identified by microscopy alone, it is abundantly clear that many adverse and toxic reactions in patients could have been avoided if basic procedures using microscopy had been applied. There have been numerous case reports of atropine poisoning arising from contamination of herbal teas supposedly containing Comfrey (*Symphytum*), nettles (*Urtica*) or Burdock (*Arctium*) with Deadly Nightshade (*Atropa belladonna*) leaves or roots. The so-called "Hairy-baby syndrome", which was believed to have been caused by the use of Siberian Ginseng (*Eleutherococcus senticosus*) during pregnancy, turned out to be due to the substitution of Silk Vine (*Periploca*) for the Siberian Ginseng. Microscopy would have shown this adulteration as it would have shown up the more recent adulteration of Plantain (*Plantago*) by *Digitalis lanata* (woolly foxglove) in a diet product.

Perhaps the most notorious case where microscopy would have prevented untold suffering is the "Fang Ji" case in Belgium. In 1993 a number of young Belgian women developed renal fibrosis which was attributed to a slimming preparation containing herbs, stimulants and diuretics. Unfortunately there was inadequate quality control of the two Chinese plants used and instead of *Stephania tetrandra*, another plant with the same common name "Fang Ji", was supplied. This plant was *Aristolochia fangchi* which contains aristolochic acids which are known to be both nephrotoxic and carcinogenic. By 1994 there were 70 of these cases, many of whom have now also developed malignancies. This tragedy illustrates quite clearly the problem of using common names for herbal drugs because the name "Fang Ji" applies to at least three different genera of plants and five different species of *Aristolochia*. This is why it is so vital to use the Latin botanical name on the labels of herbal medicinal products as well as the common name. This catastrophe also highlights the need for proper quality control of all medicines based on plant material because microscopy would clearly have distinguished between *Aristolochia* species with their cluster crystals of calcium oxalate and *Stephania* which has prisms of calcium oxalate.

While the first essential is to establish the identity of the plant, it cannot be assumed that correct identity alone guarantees efficacy and safety for a particular herb. The activity of the plant is due to the presence of one or more groups of phytochemicals. Various factors influence the amount of these phytochemicals in the plant.

Genetic factors

These are factors which relate to the inherent variability of individual organisms. In medicinal plants this genetic variation can be seen for example in the Opium poppy where the morphine content can vary from 3–12%. It is possible to improve the quality of the plant by altering its genetic structure. This mutation can be a natural process and use is made of it by natural selection of high-yielding strains of plant. For example, the Quinine content of Cinchona trees has been increased from an average of 5% to 15% simply by natural selection.

However, the use of natural selection is a slow process as it depends on very slow genetic mutations taking place in the plant. It is possible to speed up these changes by artificially inducing mutations and in so doing increase the number of chromosomes to give what are called polyploid forms.

The number of chromosomes normally present in a plant are arranged in pairs so that the complete set is composed of 2 haploid (N) sets and organisms containing them are termed diploid and represented as 2N. Most normal plants are diploid.

Polyploids

Where the number of chromosomes is increased to 3N, 4N etc. the organism is known as a polyploid. Polyploids arise by the multiplication of the chromosomes from a single species; these are termed as autoploids.

They can also result from the hybridisation of two individual species; these are termed alloploids. Natural polyploids exist, for example, in Peppermint (an alloploid) and in Valerian (an autoploid consisting of 2N, 4N and 8N forms). Most polyploids are, however, induced artificially by treating seed material with either colchicine or with ionising radiation. The effect of polyploidy on medicinal plants can be unpredictable and deceptive. For example, the 4N form of the plant Lobelia has a higher % of lobeline but since the resulting plant is very small the overall yield of the alkaloid is actually lower than that of the 2N form.

Chemical Races

Another form of genetic variation in medicinal plants is the existence of what are known as chemical varieties, chemotypes, chemodemes or chemical races. Arising from detailed phytochemical studies of plants, it has been noted that many plants of the same species, having similar external botanical characteristics, have totally different chemical constituents. Several important examples are known, including Cannabis in which there are two main chemical varieties, namely the drug type which is rich in the psychotropic tetrahydrocannabinol (THC) and low in the sedative Cannabidiol (CBD) and the fibre type (Hemp) which is low in THC and rich in CBD. The EU permits and subsidises the cultivation of the fibre type with the provision that it contains less than 0.3% THC. Among medicinal plants it is known that there are at least six chemotypes of Thyme, only one of which contains the antibacterial Thymol. In *Calamus* or Sweet Flag, used as a herbal remedy for stomach conditions in the USA, four chemotypes are known, only one of which does not contain the known carcinogen, isoasarone, the presence of which has led to restrictions on the sale of this plant.

Hybridization

The formation of a hybrid combines, in a single species, the desirable characteristics of two or more separate species. An example is peppermint which is a hybrid between Watermint (*Mentha aquatica*) and Spearmint (*Mentha spicata*). In medicinal plant research the usual tactic is to combine ("cross") a plant with a high % content of the desired compound with a related plant with high phytomass or bulk, thus giving a high overall yield.

Ecological factors

In addition to the genetic factors, a variety of ecological factors can influence the quality of a medicinal plant by influencing the phytomass or amount of dry matter produced, the ratio of different plant organs and the level of active substances. The ratio of organs can be important if the end product is a flower as in the case of Chamomile, as then factors such as light which could increase or decrease the amount of flowers produced by each plant could be significant.

Light is one of a number of factors usually classified under the heading 'climate'. The others include temperature, and water levels. The second major environmental factor is the soil.

Climate

Under 'field' conditions it is the location where the plant grows, either in the wild or cultivated, that determines the climate. Climatic conditions can only be scientifically studied in an artificial growth chamber called a Phytotron, where water, temperature, light and indeed soil can all be changed in a programmed manner.

Light

Since solar energy is the major requirement for the plant to perform primary metabolism through the photosynthetic reactions, there is an indirect effect on secondary metabolism. In general, increased sunlight increases the level of active compounds. For example, alkaloid levels in *Atropa belladonna* are 6–8 times higher in plants grown in direct sunlight compared to plants grown in the shade.

Temperature

Increases in ambient temperature can, by increasing general metabolism, result in an increase in the level of active constituent. Excessively high temperatures can cause loss of volatile materials from leaf surfaces. In the case of the fixed oil from linseed, plants grown in cooler northern climates tend to have a high level of unsaturated fatty acids while those grown in warmer areas tend to be low in unsaturates.

Water

Water is made available to plants either as rain, dew or humidity and artificially through irrigation. The effect of water is variable; many plants, e.g. cinchona, require high humidity for growth. Excessive rainfall on the other hand may encourage rotting of harvested material.

Since climate cannot be standardised in the real world, it follows that climatic variations from one year to the next, and from one area to another, can result in significant variation which must be monitored through chemical or chromatographic assays for the presence of those phytochemicals known to have therapeutic effects or which could be used as markers for the quality of the plants.

Soil

Soil is the fundamental basis of crop production, providing nutrients, water and air. Equally it should be free from harmful concentrations of pests and toxic substances. Medicinal plant crops are similar to food crops in their susceptibility to fungal and insect pests. Fungal disease can result in a 10% decomposition of Opium alkaloids in just 24 hours and fungal blight on *Solanum laciniatum* (cultivated as a source of the steroid precursor solasodine) results in a 20% loss of solasodine.

Soil chemistry

It is generally believed that soil chemistry is the most important factor influencing the yield of secondary metabolites simply because it is amenable to alteration by man. The chemicals necessary for plant growth can be subdivided into:

(a) Macronutrients
(b) Micronutrients

Macronutrients include N, K, PO_3 and lime and are usually applied to the soil as natural or artificial fertilisers. The effects of additional fertilisers on medicinal plant productivity is generally beneficial, but scientific results have often revealed contradictory results and each plant must be investigated separately to determine the optimum level and frequency of fertiliser application.

Micronutrients are elements such as Boron, Zinc, Copper, Cobalt, Iron, Magnesium etc. which are required in trace amounts. The addition of extra trace elements has no effect on plant growth but a deficiency of any or all of these can cause serious disturbance of plant growth, leading to a loss of active chemical. Since plants can be cultivated in areas with different soils of varying fertility, it follows that the site of cultivation could influence quality, indicating again the need for assays for chemical content.

Harvesting of medicinal plants

The time and method of collection of a particular plant depends on the content of its constituents and on the morphology of the plant. It is important to be able to continuously monitor the content of the plant during the growing season to ensure that harvesting is carried out at the optimum time. This monitoring usually involves chemical or chromatographic assays. More recently, extensive use has been made of radio-immunoassay (RIA) and enzyme immune assays for this purpose. Some medicinal plants are collected from the wild - so-called 'wildcrafted herbs' because cultivation is either uneconomic or technically difficult. The ability to monitor chemical content so as to ensure that the correct plant and part of the plant with optimum phytochemical composition is collected is a major advantage of cultivation.

Ontogenetic variation

One of the objectives of monitoring chemical content is the elimination of the effects of changes in chemical content which occur in the plant during the growing season, referred to as ontogenetic variation. Examples include Peppermint where young leaves contain the toxic monoterpene pulegone whereas the more desirable menthol and menthone are found in the older leaves. Similarly, the morphine content of the Opium poppy is at its highest 2-3 weeks after flowering. Conversely the solasodine content of various Solanum berries actually disappears as the berry ripens. In general terms, leaf drugs are collected as the flowers begin to open, flowers just before they are fully expanded and underground organs (roots, rhizomes) etc. just as the aerial parts die down.

Drying and preservation of plant drugs

When the plant has been harvested, various biochemical processes continue for some time, as the enzymes of the cell continue to function normally and abnormally. The cells die slowly and enzyme activity is no longer under any control. This phenomenon is referred to as post harvest enzyme activity and it can be either desirable or undesirable in its consequences.

Desirable post harvest enzyme activity

Many plants owe their activity to compounds which are only formed as a result of post harvest enzyme effects. Examples include vanilla pods where there is no vanillin in the freshly picked pod; it is formed only during fermentation of the pod which involves an enzymatic hydrolysis of a glycoside.

Plants such as Wild Cherry bark and Bitter almonds owe their medicinal use to the formation of HCN and benzaldehyde as a result of enzyme action on the cyanogenetic glycosides.

The major laxative constituents of Senna Leaf and Pod are the sennosides and it has recently been established using RIA that the sennosides are only formed when the leaf and pod are allowed to dry gradually at room temperatures.

Undesirable post harvest enzyme activity

This can involve the loss of active constituents, e.g. opium can lose up to 50% of its morphine content because of a peroxidase, or it can involve the formation of undesirable constituents. Castor Oil is an example of this because the freshly prepared oil contains a lipase which gives rise to free fatty acids which are undesirable in the oil.

Prevention of post harvest enzyme activity

In drug plants where enzyme activity is damaging, the activity can be prevented by drying because many of the most damaging enzymes are hydrolases requiring moisture for activity. Most plants contain up to 90%

by weight of water which is the basic medium for all the biochemical reactions in the plant. Therefore the elimination of this water should ensure that these reactions do not take place. Rapid drying can also ensure that proteinaceous enzymes are denatured, as well as preventing decay and the growth of microorganisms. A water content of 5–10% is generally too low to support either enzyme action or microbial growth.

Storage of crude drugs

Generally speaking, storage of medicinal plants is undesirable but largely unavoidable. This is because dried drugs are extremely hygroscopic and can absorb up to 15% by weight of moisture from the atmosphere. Such drugs are termed "air dry" and are susceptible to a variety of changes involving:

1. Enzymatic transformations: Dried plant drugs almost always contain some enzyme content which has not been destroyed during drying. In the presence of moisture this enzyme may be reactivated with harmful consequences to the chemical components of the plant.
2. Oxidation reactions: affecting volatile oils particularly.
3. Rancidification: affecting lipids.
4. Photochemical changes.

Many natural products are unstable in light, particularly the Indole alkaloids such as Reserpine from Rauwolfia and the Ergot alkaloids, Ergotamine and Ergometrine. Obviously such products need to be protected from light even during quality control procedures. The non-nitrogenous cannabinoids from cannabis are also unstable in light, and THC, the major component, virtually disappears after two years in carelessly stored drug.

Environmental contamination

With the increasing popularity of phytomedicines (i.e. medicines based on plants) both in conventional and complementary medicine, there has been increasing concern over the health impact of environmental contamination of vegetable drugs. Such contamination can take a number of forms:

1. Microbial
2. Insect and Rodent Pests
3. Heavy Metals
4. Radionuclides
5. Pesticides

Microbial

Both bacteria and fungi can contaminate plant drugs. Bacterial contamination, particularly with enterobacteria e.g. *E. coli*, is often associated with contamination by animal fertiliser or human excreta. Such contamination puts patients at risk from infection. This is especially so for patients who are immunosuppressed or who are elderly or very young because pathogenic organisms such as *Staph. aureus*, *E. coli*, *Salmonella*, *Shigella*, *Pseudomonas aeruginosa* and *Listeria* have been encountered. Many herbs have levels of contamination similar to those found in fruit and vegetables. The use of alcohol to extract plants or boiling water to produce infusions (teas or tisanes) does significantly reduce contamination although it is important that such teas are not kept for more than 24 hours. Where cold water is used for extraction, no decontamination occurs and each dose must be freshly prepared. Industrially a number of methods are available to reduce or eliminate bacterial contamination. These present a problem for organic growers of medicinal plants who have a particular difficulty since they use natural manures yet at the same time many decontamination methods are contradictory to the spirit and philosophy of their 'Green' approach to cultivation.

The European Pharmacopoeia has published guidance on the microbiological quality of Pharmaceutical preparations which includes Herbal Remedies consisting of one or more whole or powdered vegetable drugs. The limits set distinguish between Herbal Remedies in the form of teas to which boiling water is added and other Herbal Remedies. In the former case the total viable count (TVC) for aerobic bacteria is not more than 10^7 bacteria per gram of herb. In addition the presence of *E. coli* is limited to 10^2 per gram. In all other cases the requirements are stricter, reflecting the view that the boiling water can reduce bacterial loads. These other herbs require a TVC of not more than 10^5 aerobic bacteria but more importantly enterobacteria are limited to 10^3 per gram and both *E.coli* and *Salmonella* must be totally absent. Surveys indicate that most products tested meet the Guidance limits although a significant number of preparations for oral or topical use have exceeded the limits, as have samples of herbs themselves.

A variety of white, blue and green moulds can be found contaminating plant drugs. The fungal mycelia penetrate the bulk of the drug, producing an undesirable appearance and odour. It is also likely that the active principles may decompose. Some fungal transformations, e.g. in coumarin-containing herbs, may lead to the production of anticoagulant dicoumarol-type dimers. Even more worrying is the fact that many plant drugs are imported from tropical and semi-tropical areas and because of the high temperatures and humidity, some of them have been found to contain fungi capable of producing mycotoxins such as the mutagenic and teratogenic aflatoxins which also cause liver cancer. The European Pharmacopoeia suggests that tests for aflatoxins be carried out. Bacterial endotoxins could be a problem in the small number of phytomedicines based on *Echinacea, Dionaea* or *Viscum* designed for parenteral use.

Insect and rodent pests

Insects

Insects such as mites, beetles and moths can attach any stored crude drug in the same way as they attack stored food. It is the larval stage which causes the damage to the product.

Rodents

Stored plant drugs are often found to be contaminated by rat and mice hairs, urine and faeces. Infested drugs are not recognised as official by the British Pharmacopoeia which states in the General Notices that "Vegetable drugs are required to be free from insects and other animal matter and from animal excreta". Insects and rodent hairs (an indicator of rodent faecal contamination) can be recovered by preferential wetting of their exoskeletons by paraffin. Microscopical examination can establish the type of insect or rodent involved.

Prevention and control

Pests, once established, are extremely difficult to eradicate and prevention is easier and less expensive than the total sacrifice of the contaminated product. Apart from the commonsense hygiene precautions, many commercial firms use a variety of techniques to eradicate insects in stored crude drugs. None of the methods presently available are ideal: dry or moist heat may damage the constituents; fumigation with hydrogen cyanide or methylbromide may cause toxicity, while the commonly used ethylene oxide has been banned by the EU because of the suspected carcinogenicity of the chlorohydrin formed from the gas. Some companies use irradiation. Many studies have been conducted on the effects of this controversial method on the constituents of medicinal plants. Most reported studies show no effect on chemical content, with the exception of opium where even low doses of radiation reduced the morphine content. Newer experimental methods include the use of pressurised CO_2 and the use of microwaves. In one study with Senna leaves, microwaves did not appear to affect the stability of senna and various test insects incorporated into the drug sample were effectively eliminated. The European Pharmacopoeia requires that manufacturers of herbal products demonstrate that any decontamination treatment used by them does not affect the plant's constituents and that no harmful residues remain.

Heavy metals

The EU guidelines requiring testing for heavy metals in herbal remedies reflect increased contamination of the general environment and also the use of exotic Asian remedies which are heavily laced with arsenic and lead. Among other metals of concern are mercury, cadmium and thallium. Sources of these elements vary from industrial pollutants and organomercurial fungicides to atmospheric lead from petrol exhausts. In general, many different herbs have been found to exceed set limits.

Radionuclides

Radioisotope contamination of medicinal plants became a significant problem following the Chernobyl accident in 1986. In particular there was and still is concern about levels of Cs-134 and Cs-137 in herbs from Eastern Europe.

Pesticides

As a result of the increasing use of pesticides on drug crops there is continuing concern over the presence of residues in crude drugs. There is evidence that many plants have residue levels higher than those permitted in foods, e.g. DDT levels of 13.2 mg/kg have been found in Raspberry leaves and 10.6 mg/kg in Frangula bark. The European Pharmacopoeia has established limits for a listed range of organochlorine and organophosphate pesticides in vegetable drugs. For example, the limit is 1 mg/kg drug for DDT (an organochlorine) and malathion (an organophosphate).

The term pesticide includes chemicals used to eradicate rodents (rodenticides), fungi (fungicides), weeds (herbicides) and insects (insecticides). These agrochemicals may be used directly on the medicinal plant crop itself, they may be used on crops growing adjacent to the herbs or they may occur as general environmental pollutants in soil, air or water. The presence of insecticide residues is of particular concern because those of the organochlorine type (DDT etc.) have been shown to cause cancers in animals and those of the organophosphate and carbamate types are potent cholinesterase inhibitors.

For pesticides not included in the Pharmacopoeial list, a formula based on the Acceptable Daily Intake (ADI) for the particular chemicals, as published by the Food and Agriculture Organisation (FAO-WHO), is used to determine the permitted level. The Pharmacopoeia recognises that many pesticides are poorly soluble and that an extraction method used to produce extracts, tinctures etc. can modify the content of pesticides in the finished product. In such cases the calculation of the limit includes an extraction factor.

Analytical methodology

The variability in the content of phytochemicals in medicinal plants described in this chapter requires monitoring using modern quality control and quality assurance methodologies. A variety of chemical and chromatographic techniques must be used in conjunction with macroscopical and microscopical methods to ensure quality. Chemical techniques include qualitative colour tests for key groups of compounds, e.g. alkaloids. Where there is knowledge of the constituents responsible for the biological activity of the plant or where suitable marker compounds are available then quantitative assays based on titration methods, colorimetry or UV spectrophotometry can be devised. Chromatographic techniques are indispensable in ensuring adequate quality in medicinal plants and their preparations. The most widely used techniques are Thin Layer (TLC), Gas (GC) and High Performance Liquid (HPLC) Chromatography although newer methods, including Droplet Counter Current (DCCC) and Supercritical Fluid (SCFC) chromatography as well as capillary electrophoresis (CE), are finding increased use. The chromatographic methods can be used qualitatively to produce "fingerprint" chromatograms or "chromatographic profiles" as they are referred to in some Pharmacopoeias, which help authenticate the plant or its extracts. HPLC and GC are particularly useful for quantitative analysis of key components of plant extracts although TLC can also be used quantitatively either indirectly by combining it with spectrophotometric analysis of compounds removed from the plate or directly by means of a photodensitometer which scans compounds on the plate. Chromatographic methods are also used to detect some forms of environmental contamination. For example, Gas Chromatography using capillary columns linked to highly specific and sensitive nitrogen/phosphorous or electron capture detectors is specified by the European Pharmacopoeia for pesticides. Heavy metal contamination can be measured using Atomic Absorption spectrometry, while standard microbiological techniques are used to determine the extent of microbial contamination in medicinal plants.

Good agricultural practice (GAP)

While pharmaceutical analysis methodology and techniques can provide information about the quality of herbal medicinal products, technology alone cannot build quality into such products. In the same way, the various regulatory guidelines and explanatory memoranda such as those on "Products of Herbal Origin" which appeared in Pharm Europa in late 1997 can set out the types of analyses, required for medicinal plant quality control and also the rationale for the choice of analyses, but they do not address the initial causes of quality defects at the plant stage, but rather address the issue at the product stage. It is vital that quality assurance and through it, patient safety, becomes the norm from the time the plant starts to grow right through until the formulated product is used by the patient.

In order to address this point the European Herb Growers and Producers Association (Europam) have developed guidelines for Good Agricultural Practice (GAP) of Medicinal and Aromatic Plants.

The document covers the starting materials, i.e. seeds and propagation material, cultivation, harvesting, drying, packaging, storage, transport, personnel, equipment and documentation so as to ensure that plants are "produced hygienically in order to reduce microbiological load to a minimum" and that they are "produced with care so that the negative impacts affecting plants during cultivation, processing and strage can be maximally limited".

The driving force behind the guidelines is to minimise low quality medicinal, aromatic and tea products through prevention. The Guidelines allow growers to improve quality and to document their procedures in a way that is consistent with pharmaceutical quality assurance norms.

The Guidelines have been endorsed by the Working Group on Herbal Medicinal Products set up by the European Agency for the evaluation of medicinal products (EMEA). That working group has proposed that a period of 5 years should be allowed for producers to implement the requirements of this particular guideline.

Reading list

1. European Pharmacopoeia, European Pharmacopoeia Commission, Council of Europe, Strasbourg.
2. Products of Herbal Origin - note on the Monograph. PharmEuropa 9 (4), 623-625. 1997
3. British Herbal Pharmacopoeia, British Herbal Medicine Association, Surrey, UK, 1996.
4. The Rules Governing Medicinal Products in the European Union. Volume 3(a) Quality. Volume 4 Good Manufacturing Practices. Annex 7 Manufacture of Herbal Medicinal Products. European Commission Brussels 1997.
5. The European Agency for the Evaluation of Medicinal Products (EMEA) Web site: *http://www.eudra.org/emea.html* for details of revisions to (4) above under the heading EMEA/HMPWG.
6. Good Agricultural Practice (GAP) Guidelines, EUROPAM 1998.
7. D. Palevitch, Agronomy Applied to Medicinal Plant Conservation. in *Conservation of Medicinal Plants.* Edited by Akerele O., Heywood V.H. and Synge H. Cambridge University Press. Cambridge 1991 p167-178.
8. *Atlas of Microscopy of Medicinal Plants, Culinary Herbs and Spices.* Snowdon D.W. and Jackson B.P. Belhaven Press, London 1990.
9. *Plant Drug Analysis* 2nd Edition. Wagner H. and Bladt S. Springer, Berlin 1996.
10. *Trease and Evan's Pharmacognosy.* 14th Edition. Evans W.C., Saunders, London 1996.
11. Zhu M. and Phillipson J.D. Hong Kong Samples of the Traditional Chinese Medicine "Fang Ji" contain Aritstolochic Acid Toxins. *Int. J. of Pharmacognosy.* 34:283-289. 1996.
12. P.A.G.M. De Smet."Toxicological Outlook on the Quality Assurance of Herbal Remedies". in *Adverse Effects of Herbal Drugs* Vol. 1, deSmet *et al.*, Editors. Springer Verlag Berlin 1992 p. 1-72.

CHAPTER IV
PLANTS AS PHYTOCHEMICAL LABORATORIES

Pure natural substances

In the therapeutic use of natural products with mild activity a crude drug and/or extract can be adequate, but in the case of a drug with a stronger effect, a pure substance is to be preferred.

The advantage of a pure substance is the exact dosage, giving a defined effect, which is not influenced by other similar substances in the plant drug. Furthermore, the pure natural substance is easier to analyse than a crude drug.

The different naturally occurring pure substances originate from photosynthesis and related biosynthetic processes in the plant.

Photosynthesis

Photosynthesis is the process which transforms solar energy into chemical energy that can be used by living cells. It is the basis of all life on earth.

Photosynthesis takes place mainly in the green leaves of terrestrial plants, but it occurs also in marine plants. The plants take carbon dioxide from the air and water is transported from the root to the leaves.

In higher plants photosynthesis takes place in the chloroplasts, a special type of plastid occurring in the leaves of Gymnosperms and Angiosperms. The chloroplasts contain enzymes and pigments, above all chlorophyll. The chemical structure of chlorophyll with conjugated double bonds enables the plant to absorb light energy and together with water, transforms this to chemical energy as energy-rich substances, primarily glucose which then can be used by the plant for the biosynthesis of pharmacologically active substances. As a byproduct in photosynthesis, oxygen is formed.

In photosynthesis there are two types of processes. The first are light reactions, divided into photosystems I and II, which are directly dependent on solar energy; the others are dark reactions, which can take place without light. In the dark reactions, the products of the light reactions, ATP and NaDPH-ase, are used for the reduction of CO_2 to form carbohydrates. The dark reactions have been mapped in great detail.

In all cells of the plant, respiration also takes place, which is the reverse of the photosynthetic process, i.e. the cells absorb oxygen and excrete carbon dioxide. During the day this process is totally balanced by photosynthesis, with liberation of large amounts of oxygen.

Nitrogen is absorbed as nitrate dissolved in water, through the root. The nitrogen-binding bacteria, which live symbiotically in the roots of Leguminosae plants, are able to absorb nitrogen directly from the air. Nitrogen is then transformed to ammonia in the plant and incorporated into the amino acids, and biosynthesised to peptides and alkaloids.

Biosynthetic pathways

The general biosynthetic pathways are illustrated schematically in figure 1. The starting point is photosynthesis, which gives rise to carbohydrates having a mechanical function in the cell walls in the form of cellulose and a biochemical function by releasing energy via glycolysis and the citric acid cycle, which leads to other biosynthetic pathways: carbohydrates are thus the source of carbon for the synthesis of organic compounds in plants.

As seen in figure 1, carbohydrates are degraded to pyruvic acid, which is oxidised to acetate in a form which can condense to form fatty acids and polyketides (includes aromatic benzenoid molecules). Fatty acids react with glycerol to give fats or lipids. A different biosynthetic pathway leads from acetate via the condensation product mevalonic acid to mono-, sesqui- and diterpenes which are ingredients in essential oils, and also to triterpenes, tetraterpenes and steroids. Acetic acid can also be transformed to amino acids via the citric acid cycle and a source of nitrogen. These can also be formed from pyruvic acid and a source of nitrogen. The nitrogen required for amino acid biosynthesis comes originally from atmospheric nitrogen which is reduced to ammonia by nitrogen fixing bacteria, thus making it accessible for the biosynthetic processes taking place in higher plants. Amino acids may then be attached to a terpene to form some alkaloids.

From carbohydrate precursors there is another biosynthetic pathway to shikimic acid and further to gallic acid and tannins. Shikimic acid is the starting material, via chorismic acid (formed by reaction with a second molecule of pyruvic acid) which gives rise to aromatic, i.e. phenolic, amino acids.

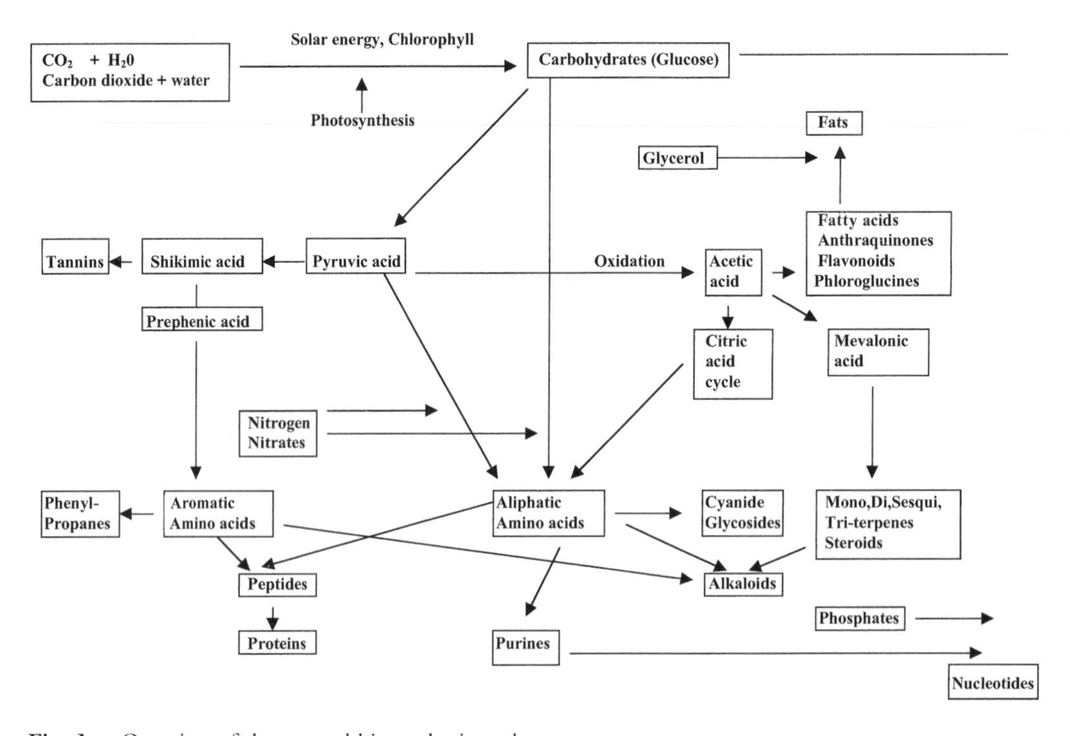

Fig. 1. Overview of the general biosynthetic pathways

The aromatic amino acid phenylalanine can also be transformed into phenyl propane derivatives, which are found in essential oils and in lignans.

An important biosynthetic pathway leads from amino acids to proteins, alkaloids, and purines, and the latter, in combination with carbohydrates and inorganic phosphate, form the vital nucleic acids.

Carbohydrates, fat and protein used to be characterised as primary metabolites, whereas the metabolites formed by various side reactions are called secondary metabolites, which are of different importance for the plant.

As a general rule it is true that the more biosynthetic steps involved in the formation of a secondary metabolite the smaller the distribution that compound has in the plant kingdom. Today more than 100,000 secondary metabolites are structurally known, and more than 3,000 substances are added each year.

Secondary metabolites, which are more or less specific for a plant species, can be classified, based on their biosynthetic origin.

Pharmacologically active natural products

In this section various chemical groups of substances, which are used as pharmacologically active substances, are described.

Carbohydrates

Carbohydrates can be grouped into *sugars* and *polysaccharides*. The sugars are water soluble, more or less sweet in taste, either *monosaccharides*, such as glucose, fructose and fucose, or *oligosaccharides*, containing up to five or six monosaccharides units, which may be the same or different. Thus, *disaccharides* consist of two monosaccharide residues bound to each other, such as sucrose, (glucose and fructose); *trisaccharides* contain three such residues, and so on. Monosaccharides can be classified according to the number of carbon atoms they contain. Thus, *trioses, tetroses, pentoses, hexoses* (the most common) and *heptoses* are C3 to C7 compounds.

Sugar alcohols, such as sorbitol and mannitol, are formed by reduction of the aldehyde-group in the monosaccharide.

Polysaccharides are macromolecules, containing a large number of monosaccharide residues. Their solubility in water is very low and they have no taste. Starch, cellulose, inulin and dextran are typical polysaccharides.

Another group of polysaccharides consists of gums and mucilages, which do not dissolve in water but swell. On hydrolysis they give both pentoses and hexoses and also uronic acids, which are formed by oxidation of the aldehyde function of the basic sugar. These are also referred to as non-starch polysaccharides, or soluble fibres. See chapter VII for more details.

Glycosides

Glycosides consist of a sugar residue covalently bound to a non-sugar molecule, called the aglycone: The sugar residue is in its cyclic form and the point of attachment is the hydroxyl group of the hemiacetal function. The most common type of glycosidic link involves oxygen (O-glycoside), but it can also involve sulphur (S-glycoside), nitrogen (N-glycoside) or carbon (C-glycoside). α-glycosides and β-glycosides are distinguished by the configuration of the hemiacetal hydroxyl group. The majority of naturally occurring glycosides are β-glycosides, whereas most polysaccharides have β-glycosidic links.

O-glycosides are the most abundant. The biosynthesis of glycosides involves both the formation of the aglycone and the coupling of aglycone to the sugar. The structure of the aglycone determines its biosynthesis, but the linkage of the sugar and aglycone is always performed in the same way, regardless of the structure of the aglycone.

The glycosides are an important group of phytopharmaceuticals, the pharmacological properties of which are mainly determined by the structure of the aglycone, modified to larger or lesser extent by the sugar part. The glycosides are found in several therapeutic groups.

Examples of glycosides of various groups

Glycoside group	***Glycoside***	***Aglycone***	***Sugar***	***Occurrence***
Cyanogenetic	Amygdalin	Benzaldehyde	Glucose	Bitter almonds
Saponins	Glycyrrhizin	Glycyrrhetic acid	Glucuronic acid	Liquorice root
Cardiac	Digitoxin	Digitoxigenin	Digitoxose	Foxglove
Anthraquinone	Sennoside A	Rheindianthrone	Glucose	Senna leaf
Flavonoid	Rutin	Quercetin	Rhamnose	Buckwheat
Phenol glycoside	Arbutin	Hydroquinone	Glucose	Uva ursi leaves
Phenol glycoside	Salicin	Saligenin	Glucose	Willow

Tannins

Tannins are a phytochemically heterogenous group of natural products, the most important common property of which is their traditional ability to form insoluble compounds with proteins. This reaction is the basis for their extensive use in leather production. The same property is the reason for their traditional therapeutic use, which includes the treatment of diarrhoea, bleeding gums, and skin injuries through an astringent effect.

Tannins can be divided into two groups: *hydrolysable tannins*, which are split into simpler molecules on treatment with acids or enzymes, and *condensed tannins* (catechin tannins), which give complex insoluble products on similar treatment. Hydrolysable tannins can be subdivided into *gallotannins*, which yield sugar and gallic acid on hydrolysis, and *ellagitannins*, which furnish not only sugar and gallic acid, but also ellagic acid. Condensed tannins are complex polymers, whose chemistry is only partly known. Their components include catechins and flavonoids, which are often esterified with gallic acid.

The tannins occur in varying amounts in practically all plants. Tannins are often present in unripe fruits, but disappear during ripening. The function of tannins is not known with certainty, but it seems possible that they give protection against microorganisms.

Volatile oils

Volatile oils (essential oils) are very complex, pleasant smelling, volatile mixtures containing many different compounds. Most of the essential oils have a high refractive index and they are often optically active. These properties are used for their identification and quality control. Essential oils have low water solubility but are readily soluble in organic solvents. Chemically and biosynthetically they are a heterogenous group.

In one group of essential oils the phenylpropane derivatives (from shikimic acid) dominate; in another group mono-, sesqui- and diterpenes are the dominant substances.

Essential oils occur in plants in specially developed organs of various kinds, e.g. glandular hairs on leaves, stems and flowers. Secretory ducts or cavities (schizogenous or lysigenous) or oil cells may also be present in the plant tissue.

Plants that are rich in essential oils (0.01–10% dry weight) are found in about 30% of plant families. A high content of essential oils is especially common in the families *Apiaceae, Lamiaceae, Lauraceae, Myrtaceae, Pinaceae, Rutaceae* and *Zingiberaceae*.

Some medicinal plants and spices have no smell when they are fresh, e.g. mustard seed, valerian root and vanilla fruit. On storage or fermentation, when a hydrolysis of the glycoside (the bound form of the essential oil) takes place, the characteristic smell appears. Similar conditions are also present in anise and fennel, the essential oils of which are partly bound as glycosides.

The content and composition of the volatile oil from one species is dependent on the habitat, soil, climate, vegetation period, and sunlight. Defective drying and storage can diminish the content of volatile oil.

Volatile (essential) oils can be obtained in principally four different ways:

1. Extraction (enfleurage method). A method, used since classical antiquity, where the fresh plant material (often flowers, e.g. jasmine flowers, passion flower) is introduced into fat, which is spread over large plates. After some days the essential oil in the fat is extracted with alcohol. Short extraction of the fat at 50–80°C is used for flowers of roses, hyacinths and carnations.

2. Extraction with solvent. This method involves extraction with petroleum ether at 50°C in a big soxhlet apparatus. Often the essential oil is mixed with waxlike ingredients and has a waxlike consistency, and is further purified with absolute alcohol.

3. Pressing and other mechanical methods. This method is mainly used in food production of citrus-fruit oils. According to the sponge method, the lemon peel is smoothly pressed to rupture the oil cavities of the peel and liberate the oil, which is absorbed in sponges, which are then squeezed out. In the Ecuelle method, machines are used to prick or cut the outermost layer of the peel, taking care not to touch the zest, since this contains enzymes that can degrade the constituents of the oil.

4. Distillation method. There are several different methods:

a) Water-distillation The fresh drug is totally covered with water in the distillation apparatus, then it is heated and distilled, e.g. turpentine.

b) Water-steam-distillation The fresh drug is firstly macerated with water, then steam is passed through the plant material. This method is used for less sensitive volatile oils, like Eucalyptus, Peppermint, Anise, Fennel and Cinnamon.

c) Watersteam-distillation The fresh plant material is put on a grating and steam is passed through under pressure from below. On passing through the plant material, the steam takes the volatile oil along with it, and the oil separates from the water upon condensation. Steam distillation is a simple method that can be used on a large scale. However, it is not applicable if the volatile oil contains readily hydrolysable substances, e.g. esters, or substances that are oxidised or undergo other changes at raised temperature. From the quality viewpoint essential oils obtained by distillation are inferior to those obtained by extraction.

The essential oils and plant materials containing such oils are natural products of great economic importance. The main use is in the perfumery and cosmetic industry and in food processing as spices. A few essential oils or pure substances isolated from such oils are used in pharmacy for several purposes:

1. Rubefacient: To increase the blood circulation in the skin, giving a feeling of warmth in cases of muscular pains: e.g.

turpentine, camphor, mustard oil, rosemary oil.

2. ***Antiinflammatory:*** Chamomile

3. ***Expectorants:*** Thyme, anise, fennel, eucalyptus.

4. ***Spice and appetite stimulating:*** chilli, mustard, pepper, nutmeg, ginger, cardamon, cinnamon and saffron.

5. ***Anti-flatulence, spasmolytic:*** chamomile, peppermint, caraway, coriander.

6. ***Diuretic:*** Juniper berry, parsley.

7. ***Emmenagogue (Menses – increasing):*** Apiole from parsley root, juniper oil, savin oil.

8. ***Nervina:*** Valerian, melissa, lavender.

9. ***Disinfectives:*** Myrrh, eucalyptus, salvia, citronella.

10. ***Anthelmintics:*** Chenopodium oil (ascaridol).

11. ***Corrigentia for smell and taste:*** lemon oil, bitter-orange oil, bergamot oil, rose oil.

Resins and balsams

A balsam is a solution of resin in a volatile oil and is generally produced by a special cell in the plant. It has a syrupy consistence. The formation of balsam can be stimulated by wounding the wood.

When a balsam is distilled, a volatile oil and a resin are obtained. This latter is an amorphous mass, which is softened by heating and melts to a clear, viscous and sticky liquid. A typical balsam is biosynthetically related to the terpenes and phenylpropane derivatives of volatile oils. Balsams are found in several plant families, e.g. *Anacardiaceae, Apiaceae, Burseraceae, Fabaceae, Pinaceae, Styracaceae.*

Bitters

Bitters (amara) are usually not isolated compounds, but aqueous or ethanolic extracts of crude drugs containing bitter-tasting but otherwise pharmacologically relatively inert substances, exclusively used for stimulation of production of saliva, gastric juice and bile. Thus, substances with bitter taste which have other pronounced pharmacological properties, like aloe, cardiac glycosides and Nux Vomica seed, are not included in this group. The bitter taste is modified by other substances in the drug. Chemically one can differentiate between terpenoids (e.g. gentiopicrin from Gentian root) and nonterpenoid bitters (e.g. quinine from Cinchona bark and humulone from Hops).

Gums and mucilages

Gums and mucilages are old pharmacognostical terms for certain polysaccharide products. They are now referred to as nonstarch polysaccharides or soluble fibres. They may contain both pentose and hexose residues, as well as their oxidation products, uronic acids. The carboxyl groups are usually neutralised by calcium or magnesium ions. It is claimed that gums are normally fairly soluble, whereas mucilages do not dissolve, but swell to a viscous mass. It is further stated that gums are pathological products, but mucilages are normal plant constituents. However, it has been difficult to apply this distinction in a consistent way, since several drugs and products are intermediates between the two groups.

Gums and mucilages are produced in various ways in the plant:

A. They are formed from the middle lamella of cells. This is common in algae, e.g. agar.

B. They are formed from the entire cell wall, e.g. the seed-coat epidermis of flax (linseed).
C. They are formed in special mucilage-secreting cells, e.g. in squill.
D. They are formed as a product of cell-wall decomposition, e.g. tragacanth, sterculia gum, gum arabic.

Gums and mucilages have great technical use in pharmacy as tablet binders and disintegrants, as thickeners and emulsifiers. Some drugs that are insoluble in water but have a great capacity to swell are used as bulk laxatives in the same way as some cellulose derivatives, and a number also reduce cholesterol and lipid levels in plasma.

Terpenes

The basic unit in the biosynthesis of terpenes is the isoprene molecule, isoprenfenylpyrophosphate formed from acetate in the mevalonic acid pathway, which can self-condense in various ways: usually "head to tail", less frequently, "tail to tail" or "head to head".

Several different groups of terpenes are of pharmacognostical importance, as shown below:

Class of terpenes		Isoprene units	Occurrence
Monoterpenes	10C	2	Volatile oils, iridoids
Sesquiterpenes	15C	3	Volatile oils
Diterpenes	20C	4	Volatile oils, resins, Vitamin A
Triterpenes	30C	6	Squalene, pentacyclic triterpenes, steroids, cardiac glycosides
Tetraterpenes	40C	8	Carotenoids, xanthophyll
Polyterpenes	nC	n	Rubber

Steroids

Steroids are formed from two sesquiterpene molecules – farnesyl pyrophosphate – which are attached to each other "tail to tail" to give the acyclic squalene (30C-atoms); this undergoes complex reactions involving a cyclisation to tetracyclic or pentacyclic ring systems, especially cholesterol.

Cholesterol (27C) can be transformed into steroid saponins, cardiac glycosides and steroid alkaloids in plant cells and into steroid hormones in animal cells.

Saponins

Saponins are a group of glycosides which have characteristic properties, i.e. they lower the surface tension of a water solution (forming a foam) and haemolyse red blood cells.

Chemically the saponins can be divided into two groups according to the structure of the aglycones.

1. Steroid saponins have an aglycone with a steroid skeleton. They are not used therapeutically, but the aglycone is used as a starting material for the synthesis of steroidal hormones (contraceptives, corticosteroids etc.) by the pharmaceutical industry.
2. Triterpene saponins have an aglycone with a triterpene structure (e.g. primulagenin and quillajic acid). These saponins are used therapeutically as expectorants; others are antiinflammatory and others are believed to have adaptogenic properties.

The ability of saponins to form precipitates with cholesterol, which is an important component of all cell membranes, can be used in suitable dosage to increase the absorption of substances via skin or mucous membranes, e.g. drugs which are only poorly absorbed may then be absorbed via the gastrointestinal tract. Small amounts of saponin increase the absorption of calcium and other nutrients. Therefore saponin-containing plants like oats, red beets, tomatoes, mangold and spinach increase the absorption of nutrients in the gastrointestinal tract. Furthermore, the saponins stimulate the secretion of digestion enzymes. Some saponin-containing plants are of therapeutic importance, e.g. *Panax ginseng*, liquorice and horse chestnut.

Acetogenins

This group consists of fatty acids, fats and waxes, flavonoids, anthraquinones and phloroglucinol derivatives.

The fatty acids are divided into saturated and unsaturated types, based on the presence and number of double bonds.

Fats and waxes are esters of fatty acids and alcohols. Fats are esters of fatty acids and glycerol. Solid fats contain saturated fatty acids, while liquid fats (fixed oils) have a high content of unsaturated fatty acids. Fats occur in both plants and animals. In plants, fats are reserve nutrition in fruits and seeds. The pharmaceutical use of fats is as a component in ointments and liniments for external use, where they are vehicles for active molecules.

Waxes are esters between fatty acids and high molecular alcohols such as cetyl- or myricyl alcohol. Waxes occur in plants as a coating on the epidermis on leaves and fruits with the function to protect against loss of water. Also, insects such as bees produce waxes.

Flavonoids are coloured substances and occur as pigments in plants. The chemical structure is built from an acetate-derived component and a segment from the shikimic acid biosynthetic pathway. Knowledge of the pharmacological importance of flavonoids has increased in recent years, including their role in reducing the risk of certain diseases (Chapter VII).

Anthraquinones are derived from the tricyclic anthracene molecule; they have one or two keto-groups in the middle ring and one phenolic OH-group on each of the outer rings and have a strong laxative effect. Phloroglucinol-derivatives occur in ferns and have one, two or three rings, connected via methylene groups: the greater the number of rings, the stronger the anthelmintic effect. It is claimed that the phloroglucinol derivative hyperforin is a major active molecule in St. Johns wort (*Hypericum perforatum*).

Alkaloids

Alkaloids, which are among the most important groups of pharmacologically active phytochemicals, usually are defined as basic, nitrogen-containing substances from plants or animals with more or less pronounced pharmacological effects. The distinction between other nitrogen-containing substances within the plant kingdom is more a matter of convention than chemical difference.

It has been proposed from a biosynthetic point of view to divide alkaloids into protoalkaloids, pseudoalkaloids and true alkaloids.

Protoalkaloids are bases with a simple structure, e.g. the biogenic amines, which are formed by decarboxylation or N-methylation of an amino acid (tyramine, histamine, choline, ephedrine, mescaline). Capsaicin, which is the amide of vanillylamine (4-hydroxy-3-methoxybenzylamine), has also been included in this group.

Pseudoalkaloids are bases, the basic structure of which is connected to a non-alkaloid component, and its nitrogen is a secondary phenomenon, e.g. steroid alkaloids and diterpene-alkaloids.

The 'true' alkaloids contain the nitrogen bound in heterocyclic form. Their carbon

skeleton is generally formed by attaching an aminoacid to a non-nitrogenous compound via several biosynthetic steps. Often these alkaloids contain two or more nitrogen atoms in the molecule.

Amino acids, peptides, proteins, nucleotides, porphyrins, vitamins, nitro- and nitroso compounds are not considered as alkaloids.

Some of the most important structural types of alkaloids are listed below:

Structural type	*Alkaloid*
Amino alkaloids	Ephedrine, mescaline, colchicine
Pyridine and piperidine	Coniine, arecoline, nicotine
Tropane	Hyoscyamine, cocaine
Pyrrolizidine	Senecionine
Quinolizidine	Sparteine, cytisine
Isoquinoline	Morphine, codeine, emetine
Indole	Psilocin, physostigmine, ergotamine, reserpine, strychnine
Quinoline	Quinine, cinchonine
Imidazole	Pilocarpine
Steroid	Veratrine, solasodine
Diterpene	Aconitine
Purine	Caffeine, theobromine, theophylline

The alkaloids are primary, secondary, tertiary amines or quaternary ammonium bases. Most are tertiary amines. The basicity of the alkaloids varies from strong bases like choline to very weak bases like the purine alkaloids.

In plants, the alkaloids occur as water soluble salts with plant acids, like acetic acid, oxalic acid, lactic acid, tartaric acid, citric acid. Some alkaloids are also bound to tannins.

The free bases are soluble in lipophilic organic solvents, like chloroform. The different solubility for bases and salts is used for isolation and purification. Alkaloids are generally colourless solid materials. Some alkaloids that do not contain oxygen, like coniine, nicotine and sparteine, are liquid, and berberine and chelidonine are intensely yellow. Several alkaloids, e.g. strychnine and quinine, have a very bitter taste. Alkaloids give precipitates with heavy metals like mercury and bismuth. Dragendorff's reagent is used to show the presence of alkaloids.

The occurrence of alkaloids in the plant kingdom is limited to certain families. With the exception of some fungi, e.g. ergot, and *Psilocybe*-species, they are found mainly in higher plants. Within the Gymnosperma and Pterodophyta, alkaloids are found in *Equisetum, Lycopodium* and *Ephedra* species. Among the Monocotyledonae, alkaloids are found within the families Melanthiaceae and Amaryllidaceae.

Finally, among Dicotyledonae the largest occurrence of alkaloids is found in the families Apocynaceae, Berberidaceae, Buxaceae, Chenopodiaceae, Fabaceae, Loganiaceae, Menispermaceae, Papaveraceae, Rubiaceae and Solanaceae.

Alkaloids are found in all parts of the plant, especially in bark, leaves, root and fruits. Sometimes the alkaloids are found in certain cell-types of tissues, e.g. the opium alkaloids are found specifically in the latex vessels of the poppy fruit or capsule.

Their biosynthesis, which involves a combination of the main biosynthetic pathways, is fairly well known, whereas their role in the plant is more or less unknown.

Production of pure natural substances

Most pure natural substances are obtained by isolation from the plant, e.g. alkaloids, cardiac glycosides, anthraquinones, flavonoids, coumarins, bitters, saponins, high molecular peptide hormones and enzymes.

The production of pharmaceutical natural products can also be carried out by using microorganisms, cell cultures, plant tissue and enzymes.

The microorganisms that have been used up to now on an industrial scale are bacteria and fungi, belonging the class of ascomycetes such as *Saccharomyces, Torulopsis* (yeast), *Rhizopus, Penicillium, Aspergillus.*

Only a few substances are currently produced on an industrial scale from plant tissue and cell cultures: Ergot alkaloids from the fungus *Claviceps purpurea*; in Japan two products are produced – the alkaloid berberine (for treatment of diarrhoea) from *Coptis japonica* (family Ranunculaceae) and the red pigment shikonin from *Lithospermum erythrorhizon* (family Boraginaceae). Shikonin is a remedy for skin disease and also an ingredient in lipsticks.

Up to now the use of tissue culture from higher plants for production on an industrial scale has been successful in only a few cases. On the other hand, small-scale, biotransformations of natural pure substances to new pharmacologically active compounds, e.g. cardiac glycosides, have been performed.

Chemical and microbiological transformation of pure natural substances

Natural products in the form of isolated pure substances do not always represent the optimal therapeutic effect. However, that can be modified in two ways: by chemical methods, or by microbiological methods into semisynthetic derivatives. Examples of chemical modifications are:

Chemical Modifications of Natural Products

Isolated Pure Substance	Final Product
Citronellal	Menthol
Piperitone	Thymol
Pinene	Camphor
Xanthine	Caffeine
Morphine	Codeine, ethylmorphine
Lysergic acid	Ergometrine
Strychnine	C-toxiferin
Colchicine	Demecolchine
Diosgenin	Progesterone
Glucose	Sorbitol, mannitol Ascorbic acid
Vinblastine	Vincristine
Strychnine	Strychnine-N-oxide, strychnic acid
Ergometrine	Methylergometrine
Ergotamine	Dihydroergotamine
Scopolamine	N-Butylscopolamine bromide
Emetine	2-Dehydroemetine

Chemical Modifications of Natural Products (*continued*)	
Isolated Pure Substance	Final Product
Codeine	Dehydrocodeinone
Digoxin	Methyldigoxin
Vincamine	Apovincamine
Rutin	Rutinoside
Podophyllotoxin	Etoposide, Tenoposide
Castor oil	Polyoxyethyl castor oil (solubilisation)
Gelatin	Oxypolygelatin
Starch	Oxy ethyl starch

Microbiological Modifications of Natural Products		
Isolated Pure Substance (starting material)	Microorganism	Final Products
Sucrose	*Leuconostoc* (polymerisation)	Dextrans
Sucrose	*Torulopis* (hydrolysis)	Invertose
n-Sorbitol	*Acetobacter peroxidans* (oxidation)	L-Sorbose
Glucose	*Streptomyces albus* (isomerisation)	Fructose
Starch	*Bacillus subtilis* (hydrolysis)	Maltose
Cortisone	*Corynebacterium simplex*	Prednisone
Cortisol	(dehydration)	Prednisolone

Total synthesis

A rational total synthesis is used for some simple alkaloids (adrenaline, theophylline, papaverine), low molecular peptide hormones (oxytocin, vasopressin), vitamins (biotin, folic acid, thiamine, riboflavin), phenolic acids (salicylic acid). It should, however, be observed that the synthetic substances are all racemic forms, whereas many natural products occur in the l-form.

The use of pure natural substances

The pure natural substances are used as such as remedies or as starting materials for semisynthesis, as mentioned previously. They can also be used as models for total synthesis of new remedies with improved efficiency, kinetics or safety.

Here are some examples:

Substances	New Derivatives
Cocaine	Less toxic local anaesthetics, e.g. Procaine
Tubocurarine	Shortacting muscle relaxant, e.g. Alloferin
Quinine	Specific gametoactive agent, e.g. Pyrimethamine
Adrenaline, Ephedrine	Sympathomimetics, e.g. Etilefin, Norfenefrin
Acetylcholine	Long-acting parasympathomimetics, e.g. Doryl
Dicoumarol	Long-acting anticoagulants e.g. Warfarin

Another important use of pure natural substances is as pharmacological or biochemical tools for the study of novel mechanisms of action, or of pathophysiological processes: e.g. forskolin is a diterpene in the root of *Coleus forskohlii* (Lamiaceae), which activates adenylate cyclase and causes an increase in cellular c AMP levels. Phorbol esters (from family Euphorbiaceae) stimulate protein kinase C, and tetrodotoxin in *Fugu*-species influences the transport of sodium ions.

In this chapter various types of natural products, which are produced as secondary metabolites in plants, are mentioned. It should be observed that practically all aspects of therapeutic pharmacognosy are covered with the exception of infectious diseases. Either the natural product as such is used therapeutically or as a modified structure.

For the treatment of infectious diseases, antibiotics are issued on a larger scale, but they are not dealt with here, but in microbiology textbooks and pharmaceutical courses.

Reading list

1. *Medicinal Natural Products. A biosynthetic approach*, Paul M. Dewick, John Wiley & Sons, New York 1998
2. *Drugs of Natural Origin. A Textbook of Pharmacognosy*, Gunnar Samuelsson 4th rev. Ed., Swedish Pharmaceutical Press, Stockholm 1999.

CHAPTER V
PLANTS IN COMPLEMENTARY MEDICINE

The past twenty years have seen a dramatic increase in the popularity and use of various complementary or alternative therapies. These can be referred to by many different names such as unorthodox, natural, fringe, complementary or alternative. In 1993 the British Medical Association in its report on Complementary Medicine used the term *non-conventional therapies* which it defined as '*those forms of treatment which are not widely used by the orthodox health-care professions and the skills of which are not taught as part of the undergraduate curriculum of orthodox medical and paramedical health-care courses*'.

Some of the therapies mentioned in the BMA Report are complementary in that they can be used in conjunction with orthodox treatments. For example, chiropractic, osteopathy, reflexology, Shiatsu and the Alexander technique all involve manipulation of muscle and bones and could be used to complement NSAID-based treatment. Others are genuinely alternatives which attempt to replace orthodox medicines, e.g. herbalism and homoeopathy, while some, e.g. acupuncture, are being assimilated into conventional medical practice. This chapter concentrates on those therapies which rely heavily on the use of plant materials and attempts to explain the basis of the therapy and provide examples of the plants used. In particular, the interrelationship between herbalism, herbal medicines and the phytotherapeutic use of plants, described in Chapter VI, is explored.

There are three main therapies which rely on plant-based medicines, namely Aromatherapy, Homoeopathy and Herbalism. All three, because they are plant based, exemplify the green, natural image of alternative/complementary medicine which has helped in making these techniques so popular and so acceptable. Equally the holistic approach of practitioners of these therapies, which embraces the whole person rather than seeing them as a set of symptoms or receptor-sites, has struck a sympathetic chord with patients disenchanted with the impersonal 'nature' of high-tech modern medicine. Allied to this is a sometimes naive view that 'natural' medicines are somehow free from the toxicities and side-effects of synthetic drugs. Chapter VIII shows just how erroneous such a view can be.

Aromatherapy

The medical use of volatile oils has become increasingly popular in recent years under the title Aromatherapy. This term was first used by the French chemist Gattefosse and the concept was developed further by Valnet. Aromatherapy is now a significant complementary therapy involving the use of volatile oils to heal or improve wellbeing. The oils may be administered by inhalation, oral ingestion or most usually by application to the skin by massage. Several types of aromatherapy have been described, including:

1. Classical aromatherapy used in conjunction with massage and based on the concept that the terpenoids and other constituents of the oil will be absorbed transdermally and exert a variety of pharmacological effects.
2. Clinical aromatherapy which is mainly practised by French medical doctors who use essential oils internally as alternatives to conventional medicines. Sometimes the oils are taken orally, sometimes as pessaries or as suppositories. The use of oils in this way is more likely to cause adverse reactions than when the oil is massaged into the skin. This is because massage oils are invariably diluted in a carrier or base oil such as the fixed oil from almonds, avocado, soya or peach.

There has been little scientific evaluation of aromatherapy oils but it is known that the terpenoid and phenylpropane compounds found in essential oils do possess biological activities ranging from antibacterial, antiinflammatory, sedative etc. There is evidence that molecules such as terpenoid hydrocarbons, esters etc. can be absorbed transdermally into the bloodstream. In addition, the massage element involving physical manipulation of strained tissues as well as the psychological effects of touch and of the smell of the oil, is of significance. Some trials of aromatherapy have shown benefits in intensive-care patients, in epilepsy and in endometriosis. The most popular oils are listed below with their claimed uses.

Oil	Source	Claimed uses
Cedarwood	*Cedrus atlantica* *Juniperus virginiana*	nervous stress
Chamomile	*Matricaria recutita*	insomnia, dry skin
Eucalyptus	*Eucalyptus species*	colds, sinusitis
Geranium	*Pelargonium graveolens*	menstrual, hormonal problems
Lavender	*Lavandula species*	stress, premenstrual tension, burns
Tea tree	*Melaleuca species*	antiseptic, acne
Ylang ylang	*Cananga odorata*	depression, insomnia

Many of the oils are generally safe since they are commonly used as spices and flavourings but some can cause skin and allergic reactions. Some terpenoid ketones, e.g. thujone and pulegone, can be neurotoxic and oils containing them should not be used during pregnancy, e.g. Melissa, Sage, Basil and Hyssop. Some are inadvisable in patients with hypertension, e.g. Hyssop and Sage which should also be avoided by epileptics, as should Fennel and Rosemary. Other oils such as Black Pepper, Cinnamon, Lemongrass and the Citrus oils (Bergamot, Lemon, Orange) may cause photosensitivity reactions in combination with strong sunlight or artificial UV light.

Homoeopathy

One description of Homoeopathy describes it as the treatment of patients by administering highly diluted forms of natural substances that in a healthy person would bring on symptoms similar to those the medicine is prescribed to treat. Homoeopathy was developed by the German physician Hahnemann in the 18th century. According to the theory of Classical Homoeopathy there are several key features:

1. The concept that 'like cures like' or the Law of Similars. This could mean that a patient with a fever can be cured by

administering a remedy which causes a rise in temperature. In practice, the symptom complex of each medicine is determined by the process of 'proving' in which the drug is given to healthy volunteers and all the symptoms which occur are recorded systematically. This is matched with the symptom complex of the individual ill patient based on a detailed case-history involving physical, mental and emotional symptoms. According to homoeopathic physicians, each patient will have a different symptom picture which must be matched as closely as possible to the medication picture in order to achieve a beneficial therapeutic outcome. Classical Homoeopathy is highly individualised and the holistic approach which includes consideration of personality and lifestyle is important also.

2. A second key feature is that of minimal dosages whereby the starting materials are successively diluted down to one part in 10^{-30}. The dilution process is referred to as potentisation in the belief that the more dilute the material the more potent it becomes. At each dilution step the material is shaken either manually or by machine - this is called 'succussion' and this process also plays a role in the potentisation. Not surprisingly, the use of potencies of 30c (i.e. 1 in 10^{-30}) makes it next to impossible to explain how, or indeed if, homoeopathy works, in terms of conventional medicine, pharmacology and physics. It is for this reason that homoeopathy has always been controversial, even though it is used by many doctors and many papers have been published showing clinical benefits in situations where the placebo effect alone could not be the explanation.

Clinical homoeopathy is one of a number of modifications of the basic approach, as is combination product homoeopathy. In the former case the choice of remedy is based on the major clinical symptoms rather than on the complete array. The formulae for the combination products were developed as a result of the clinical experience of practising homoeopaths; they have an extreme range of actions and are considered appropriate for the treatment of common ailments. A large number of commercially prepared homoeopathic remedies are readily available.

Homoeopathic remedies

The starting materials consist of plants, organic and inorganic chemicals and animal products. Plants are used as mother tinctures - extracts in alcohol. Various minerals, salts and synthetic materials are dissolved in alcohol and water mixtures or triturated with lactose prior to potentisation. Some animal products are derived from lower species, e.g. insects, some from healthy organs and tissues of food animals (cattle, sheep, pigs) - these are called sarcodes. Nosodes are homoeopathic preparations derived from human or animal pathogens, diseased tissues or organs, as well as decomposition products of animal organs all of which must be carefully sterilised before processing. Two European Union Directives deal with homoeopathic medicinal products; Directive 92/73/EEC deals with products for human use, while Directive 92/74/EEC deals with products for veterinary use. Under the provision of these Directives, homoeopathic medicines are subject to the same controls as other medicines in terms of good manufacturing practice, quality control, import and export but proof of efficacy is not required and a simplified registration procedure is put in place. Under the terms of such a simplified registration procedure, products must be for oral or external use only, no specific therapeutic claim may be made and the product must be so dilute that it may not contain more than one part per 10,000 of the mother tincture or other original starting material. Special labelling requirements are set out and files are required which document the origins of the homoeopathic stocks, the subsequent manufacturing steps and in-process control. Given that the end-product is so dilute it is essential that the starting material be precisely defined and identified. Various homoeopathic pharmacopoeias contain monographs which assist in the pharmaceutical quality control. These include the British Homoeopathic Pharmacopoeia (1991), and the Homoeopathic Pharmacopoeia of the United States (HPUS 1995). Some national pharmacopoeias, e.g. that of France, also include monographs on homoeopathic source materials. Good manufacturing practice, record keeping and validation of procedures is essential.

Commercial products can be recommended by homoeopaths or self-selected by patients for well-recognised conditions even though this might not be approved of by classical homoeopaths. Popular remedies would include Arnica for bruises, *Thuja* for verrucas and warts, Calcium Carbonate for rheumatism, Sulphur for indigestion, Belladonna for headache, Pulsatilla for cystitis and Bryonia for constipation.

Herbalism and herbal medicine

A significant amount of information about the therapeutic use of plants and their metabolites is presented in Chapter VI, some of which relates to the use of the whole plant or a simple extract thereof. Such preparations are usually referred to as herbal medicines or remedies or as phytomedicines. The use of the term herbal medicines implies a link with medical herbalism, a linkage which is to a large extent non-existent. Much of the present-day emphasis on herbal medicinal products results from the use of such products in conventional medicine. Medical Herbalism on the other hand is a form of alternative medicine with its own distinct philosophy and therapeutic approach and the gap between it and the use of herbal medicines is only partly bridged by a shared belief in the value of plant drugs in the treatment of illness. Herbal medicine is practised in various forms all over the world. One can differentiate between the Anglo-American approach, that of traditional Chinese medicine (TCM), of Kampo medicine in Japan and of Ayurvedic medicine in India among others.

Traditional systems of medicine

Soon after the elaboration by the World Health Organisation (WHO) of the Alma Ata Declaration of '*Health for all by the Year 2000*', it realised that approximately 80% of the world's population had no access to Western-style healthcare. Any attempt therefore to provide comprehensive healthcare depended on the utilisation of traditional systems of medicine which are almost exclusively plant based. Each continent and area has its own distinct herbal lore and practice but some have become more widely known through migration and through public interest. Many of the plants used in traditional pharmacopoeias are the subject of intense scientific investigation by the pharmaceutical industry looking for new bioactive molecules or for leads to new therapeutic approaches.

Traditional Chinese medicine

TCM as it is now known is perhaps the best known of the traditional systems. The philosophy underlying it reflects a distinct Chinese view of the world, which differs significantly from the western philosophical view. Much emphasis is placed on the importance of energy flows (Qi) allied to eight conditions involving concepts such as *yin*, *yang*, *full*, *empty*, *hot*, *cold*, *external* and *internal*. These in turn are linked to five phases and a classification of herbs into one of five tastes, namely *bitter*, *sweet*, *pungent*, *salty* or *sour*. Whatever the philosophical basis of ill-health and its prevention and treatment, the materia medica which has accumulated in TCM over several millennia is extraordinarily rich. Systematic evaluation by Chinese and by non-Chinese scientists has had an enormous impact not only in terms of affirming the validity of TCM but also in terms of drug discovery for conventional Western medicine. Among the most noteworthy developments has been the isolation of the antimalarial artemisinin from *Artemisia annua*, used for centuries to treat fevers. The polysaccharides from *Astragalus membranaceus* have been shown to have immunostimulant properties of value in cancer patients while the adaptogenic lignans of *Schisandra chnensis* are increasingly exploited. The clinical effectiveness of a multicomponent remedy for eczema has been demonstrated in standard clinical trials and seems to illustrate not just the value of synergistic combinations of plants but its necessity in certain cases.

Kampo or Kanpo medicine is the traditional Japanese medicine which is attracting increased attention as a therapy and also from Japanese pharmaceutical companies. Herbs can be classified into three classes with the lower class taken only during illness because they are considered to be the most toxic. Middle class drugs would be used to maintain health and used daily for intermittent periods of time. The highest class of drugs, the upper class, are mainly tonics and adaptogens and should be taken every day to enhance longevity.

Ayurvedic medicine is traditional Indian medicine and it has existed for 3,000 to 5,000 years. Ayurveda, because it literally means 'knowledge of life', encompasses more than just medicinal aspects as it includes psychological, cultural, religious and philosophical concepts. Much of the knowledge is included in the *Vedas* books of knowledge and in *Samhitas* which contain the science of Ayurveda. The medicinal aspect involves an holistic approach to balance and imbalance in the body and treatment aims to restore the balance of the whole body rather than suppress symptoms. Ayurvedic medicines are classified into six tastes and their effects linked to different *doshas* or body constitutions. In an emphasis common to virtually all traditional systems, there is an emphasis on the use of cleansing herbs. A number of Ayurvedic herbs have become more widely known in the West, including *Adhadoda vasica* which is effective in respiratory conditions, *Phyllanthus amarus* with its beneficial effects in liver disease and *Commiphora mukul* which has hypocholesterolaemic activity.

WHO traditional medicine programme

As part of its commitment to health for all, the WHO established a traditional medicine programme and has stated its support for the integration of traditional remedies into conventional healthcare whenever possible. As part of its commitment, WHO has published volume I of its Monographs on Selected Medicinal Plants, a second volume is nearing completion and a third volume is in preparation. The first volume includes plants from many different traditions, e.g. Garlic, *Angelica sinensis*, Hypericum, Paeonia etc. The monographs are a combination of quality standards and overviews of pharmacological, clinical and toxicological information on each plant. They have particular value because of their WHO provenance and because they make available, in English, information on such important non-Western plants as *Andrographis, Astragalus, Bupleurum, Curcuma* and *Prunus africana.*

Medical herbalism

Herbalism as it is understood and practised in the English-speaking world is influenced by several key concepts, foremost among which is the concept of a 'vital force' within the body, which inexorably pushes the individual towards a state of being healthy. This philosophy holds that when the vital force is strong that person is healthy but when the vital force is weakened, by, for example, stress, fatigue or environmental factors, then illness results. There is a major emphasis on sustaining and strengthening the vital force by using tonic and adaptogenic herbs. There is also a physiological rather than pathological emphasis in the therapeutic goals, which is linked with a view that the causes of ill-health rather than its symptoms should be treated.

As part of the herbalist's response, body chemistry is optimised by enhancing the nutrient supply to all parts of the body and by encouraging detoxification. Herbs with depurative effects, immunostimulants and those with effects on the liver would be incorporated into the prescription, as would plants which stimulate elimination. Such eliminatives affect not only the bowel and the kidney but also the skin and the lungs which in herbalism are also seen as organs of elimination. Therefore, a prescription from a medical herbalist is invariably multi-component, containing diaphoretics or sudorific herbs which promote sweating in feverish states - such sweats are seen by the herbalist as a key cleansing operation which removes pathogens or toxins. Also included are laxatives or aperients, diuretics and expectorants. External preparations might include rubefacients, which are plant materials which produce local vasodilaltion in underlying skin, and the principle behind their use is that the increased blood flow improves both the cleansing and nourishing of the affected tissues.

A further major tenet of herbal medicine is the selection and blending of a mixture of plants to suit the individual patient. Therefore different patients with apparently similar symptoms could be prescribed a different mixture of plants depending on the practitioner's assessment of their individual needs. Many commercially formulated herbal products can be heavily influenced by the herbalist's approach, as seen for example in a product for arthritis containing

Elderflower	(*Sambucus*)
Yarrow	(*Achillea*)
Poke root	(*Phytolacca*)
Burdock root	(*Arctium*)
Clivers	(*Galium*)
Poplar bark	(*Populus*)
Senna leaf	(*Cassia senna*)
Uva ursi	(*Arctostaphylos*)

Many of these herbs are diaphoretics and eliminatives. From a conventional perspective the only 'active' plant would seem to be the salicin-containing Poplar bark and the inclusion of the laxative Senna leaf seems irrational. However, its inclusion would not seem at all strange to a herbalist! In addition to multi-herb products, other commercial herbal medicines contain just single herbs - so-called 'simples'. Both types of product are largely designed for self-medication, and the lack of individualisation implicit in this would be strongly frowned upon by herbalists. However, in the developed world more patients consume herbal medicines which have been commercially manufactured compared to those who take medicines prescribed and compounded by a medical herbalist.

Herbal medicine and phytotherapy

In recent years the term phytotherapy has become widely used, sometimes to describe herbal medicine and sometimes to describe the use of plants and plant extracts to treat symptoms in conventional medicine. In the latter case the use of plants in this way can be self-medication or in some European countries it can involve prescribing by physicians. The analogy with chemotherapy is deliberate, highlighting the difference between the use of plant medicines in orthodox medicine, i.e. to treat the symptoms of certain diseases, and in the distinct and alternative form of medical treatment that is herbalism. Phytotherapy is not a new term, having been used by a French doctor (Leclerc) in the early part of the 20th century. The modern use of the term owes much to a German doctor (Weiss) who defined phytotherapy as '*the science which concerns itself with the application of plant medicines in sick humans*'.

The plant medicines referred to in the definition above have been known under a variety of names: herbal medicines, herbal remedies, phytotherapeuticals, phytopharmaca, phytopharmaceuticals and phytomedicines. Most recently the European Union through its Agency for the Evaluation of Medicines (EMEA) has decided to use the term '*herbal medicinal products*' to describe these materials. For medicine licensing purposes this type of product is defined as containing as active ingredients only plants, part of plants or plant materials or combinations thereof. Medicines containing chemically isolated plant constituents, e.g. Digoxin or Codeine, or mixtures of chemically defined substances and plant materials are therefore not considered to fall within the definition. This legalistic approach is in contrast to the concept, elaborated by Weiss, that phytotherapy embraces a sliding scale of plant drugs. At one end of the scale, one has highly potent and highly purified chemicals isolated from the Opium poppy or the Yew tree for example, which are usually reserved for medical prescription. At the other end of the scale there are whole plants with mild activity suitable for self-medication, such as chamomile or valerian.

What might be called modern phytotherapy places much emphasis on the use of whole plants or their simple ('crude') extracts with mild activity. Ironically this makes the assessment of the safety, efficacy and quality of the medicines difficult, firstly because the whole plant is phytochemically more complex and more variable than an isolated purified single compound and secondly because mild biological activity is more difficult to measure in both pharmacological assays and clinical trials. The technical difficulties encountered reinforce the doubts and scepticism of those who see the increasing popularity of phytotherapy and indeed of the alternative therapies as a retrograde step. The reluctance of healthcare professionals to accept the validity of herbal medicinal products is sometimes cloaked in claims of scientific objectivity but does not appear to be justified on objective scientific grounds, when one realises that the claim that a plant remedy acts as a result of its complex mixture of constituents rather than a single chemical has been accepted in the case of Senna and Liquorice. In the case of Senna it is accepted that the whole fruit (or pod) is preferable as a laxative because studies have shown the whole plant to be more effective than any of the sennosides isolated from the plant. In the case of liquorice (*Glycyrrhiza glabra*) it is known that the anti-ulcer effects are found in two quite distinct groups of phytochemicals, namely the triterpene saponins and the isoflavonoids. If the chemical diversity of the 'active' materials can be accepted for liquorice, it is legitimate to ask why it cannot also be accepted for chamomile for instance with its sesquiterpenes and flavonoids.

The view is frequently expressed that herbal remedies are mere placebos. There is no doubt that as with any medicine there will be a placebo response but the scientific evidence suggests that herbal medicinal products do contain pharmacologically active material. One possible conceptual difficulty might be due to the fact that the chemical compounds responsible for this activity are not necessarily the classical phytochemical groups which are well known and scientifically accepted as having bioactivity. Many herbal medicines contain so-called minor secondary metabolites, e.g. flavonoids, lignans, triterpenes etc.

These compounds often do not have the biopotency to warrant the interest of a pharmaceutical company, which might wish to develop the compound as the basis of a single chemical entity. It is vital, however, that weak biopotency should not be confused with no biopotency. An objective scientific evaluation of the claims for efficacy made in relation to many herbal products must take account of the enormous strides in phytochemical analysis, which has increased our knowledge of the constituents of medicinal plants. Equal importance must be placed on the large amount of phytopharmacological information gleaned through the application of modern receptor-binding assays for example to plant extracts. It is noteworthy that as the testing methodology becomes more sophisticated, previously neglected or obsolescent plants show activity that is scientifically unexpected but which tends to confirm traditional folkloric reputations. Examples include the impact of flavonoids on many key enzyme and receptor systems, including benzodiazepine receptors.

While phytochemical and pharmacological information is helpful, the true test of the medicinal value of a plant lies in its demonstrated clinical usefulness. The difficulties involved in submitting plants and plant extracts to rigorous clinical scrutiny in the absence of patent protection have been well documented. Equally the fact that a number of important plants have been subjected to randomised controlled trials has been documented. While some trials can be criticised in the same way that trials of synthetic drugs can be criticised, the table below shows the types of plants which have been clinically evaluated and found to be superior to placebo. For some plants there is even meta-analysis data available.

Plant drugs which have been subjected to clinical trials

Name	Botanical Name	Condition
Feverfew	*Tanacetum parthenium*	Migraine prophylaxis
Garlic	*Allium sativum*	Hypercholesterolaemia
Ginger	*Zingiber officinale*	Travel sickness
Hawthorn	*Crataegus monogyna*	Heart disease
Lemon balm	*Melissa officinalis*	Cold sores
Psyllium	*Plantago ovata*	Hypercholesterolaemia
St. Johns Wort	*Hypericum perforatum*	Depression
Valerian	*Valeriana officinale*	Insomnia

The legal status of products containing medicinal plants is somewhat confusing at the present time with different countries adopting different approaches. Quality control remains a significant issue in relation to both efficacy and safety. As outlined in Chapter III, many issues of safety of herbs are in fact due to poor quality control. However the methodology and technology is available to ensure that patients are presented with products of the highest quality. Much more controversial are the methods by which efficacy and safety can be judged. This area is complicated not only by the costs of licensing procedures but also by the attempts to incorporate centuries of traditional human experience with a given plant into the evaluation process and by the fact that some herbal medicinal products, e.g. garlic and ginger, are much more widely used in food than in medicine. A bewildering choice of food or dietary supplements is presented to the consumer by pharmacies, by health food shops, by mail order or through multilevel marketing organisations. Attempts to ensure consumer protection from fraudulent claims and dangerous products present a difficult scientific and regulatory task.

Within the United States, the most recent attempt to balance the interests of all involved is the Dietary Supplement Health and Education Act (DSHEA) which includes herbal products within the definition of a dietary supplement. DSHEA assures availability of supplements as long as they are not unsafe and allows producers to make certain structure and function claims on the label, but in words which do not apparently provide the exact information a consumer needs. In addition the DSHEA requires a disclaimer to appear on the label which indicates to the patient/consumer that the product has not been evaluated by the Food and Drug Administration and that it is not intended to diagnose, treat, cure or prevent disease.

This latest American system means that consumers must rely on third-party literature to describe the benefits and risks of a particular plant product. According to the Act, the literature must not be attached to the product and could include articles, book chapters or official abstracts of pre-reviewed scientific publications, which present a balanced view of the available scientific information. In the US at the present time great stress is placed on the information contained in the 300 or so Kommission E monographs prepared by a subcommittee (Kommission E) of the then German medicines regulatory agency, the BGA, and published in the Official Gazette prior to 1993. There is no doubt that the information in those monographs does present an accurate view of the efficacy and safety issues of plants evaluated by the Kommission, up to the date of publication. However, since Kommission E ceased publishing monographs in 1993, no updated information is available and no reference bibliography for each herb was made available.

An alternative source of authoritative information, which attempts to redress the situation, would be the WHO monographs referred to earlier. The Kommission E monographs reflect one view of the regulatory process of herbal medicinal products in a European context. Different approaches have been taken in France and Britain, for example, even though one Directive (65/65/EEC) governs the marketing of medicines for human use throughout the 15 member states of the European Union. The European Agency for the Evaluation of Medicinal Products (EMEA) has set up a Working Party on Herbal Medicinal Products. This Working Party is examining ways of evaluating herbal products for quality, safety and efficacy. In its work on the latter two aspects, the EMEA has had recourse to the WHO monographs and also the 60 monographs published by the European Scientific Cooperative on Phytotherapy (ESCOP). ESCOP is the European umbrella organisation for national associations of phytotherapy, founded in 1989. Its general aims are to advance the scientific status of what it calls phytomedicines (herbal medicinal products) and to assist with harmonisation of their regulatory status at the European level. A Scientific Committee (of which both authors are members) has compiled proposals for European monographs summarising the medicinal uses of plant drugs in the format of a Summary of Product Characteristics. The SPC format was chosen because it is an integral party of any application for an authorisation to market a medicine for human use in the EU. The information on each plant is evaluated by the multinational multidisciplinary committee and covers the pharmacopoeial definition, chemical constituents, therapeutic indications, dosage and route of administration, special warnings, precautions, contraindications and interactions and the consequences of overdosage. The major section reviews and summarises the scientific evidence on efficacy by detailing the pharmacodynamic information followed by the results of studies in humans up to and including randomised controlled trials. The section on preclinical safety data is another key element, as is the list of references to the scientific literature used in the compilation of the SPC/monograph. Plants such as Hawthorn, Ginger, Ispaghula, Senna, Hypericum, Valerian, Passiflora, Echinacea are among the harmonised monographs which have been agreed and published so far. The EMEA has adopted a number, e.g. Senna, Valerian and Ispaghula, as what are called 'Core-SPCs'. The evaluation of existing and new information on the efficacy and safety of the plants used in phytotherapy is continuing amid scientific/political efforts to ensure the availability of plant medicines to the public while maximising public health and consumer safety. A new regulatory framework, which can reconcile both objectives, is possible.

Conclusions

The value placed by the public on access to 'natural' medicines and alternative therapies continues to expand enormously. Plants provide the backbone of this phenomenon and also that of traditional systems of medicine which are essential for so many human beings. The scientific exploration of the many facets of medicinal plants, their growth, analysis, effectiveness and safety is a major endeavour involving different scientific disciplines and experimental techniques and methodologies. New medicines based on plants will be discovered, while the value and risks of old medicines will emerge from this work.

Reading list

1. *Tylers' Herbs of Choice* Robbers J.E. and Tyler V.E. The Hawthorn Herbal Press New York 1999.
2. *Adverse Effects of Herbal Drugs.* Edited by P.A.G.M. deSmet, K. Keller, R. Hänsel, R.F. Chandler. Springer Verlag Berlin, Volume 1 1992. Volume 2 1993. Volume 3 1997.
3. *Encyclopedia of Common Natural Ingredients used in Food, Drugs and Cosmetics.* 2nd Edition Leung A.Y. and Foster S. Wiley - Interscience New York 1996.
4. *Herbal Medicine* R.F. Weiss Beaconsfield Arcanum 1987.
5. *British Herbal Compendium* Vol. 1 Edited by P.R. Bradley BHMA Dorset (UK) 1992.
6. *Essential Oil Safety.* Tisserand R. and Balacs T. Churchill Livingstone. Edinburgh 1995.
7. *Herbal Medicines, A Guide for Healthcare Professionals.* Newall C.A., Anderson L.A. and Phillipson J.D. The Pharmaceutical Press, London 1996.
8. *The Complete German Commission E Monographs.* Blumenthal M. *et al.* Editors. American Botanical Council 1998.
9. *ESCOP Monographs on the Medicinal Uses of Plant Drugs.* ESCOP Exeter (UK) 1997-1999.
10. *Encyclopaedia of Herbal Medicine.* Bartram T. Grace Publishers Dorset 1995.
11. *Complementary Medicine, New Approaches to Good Practice.* British Medical Association Oxford University Press. Oxford 1993.

General Resources.

(a) *Phytochemistry*, A journal devoted to phytochemistry in general.
(b) *Journal of Natural Products* Which contains scientific studies of both plant and animal compounds with biological activity.
(c) *Planta Medica*, as the name suggests covers research into all aspects of medicinal plants.
(d) *Phytomedicine* includes clinical and toxicological research articles on herbal medicinal products.
(e) *Fitoterapia* frequently contains review articles in English on key medicinal plants.
(f) *Pharmaceutical Biology* (formerly the *International Journal of Pharmacognosy*) contains research on different aspects of plant drugs.
(g) *Herbalgram* published by the American Botanical Council is full of news items on all aspects of herbs and herbal medicines and also contains excellent review articles.

CHAPTER VI
THERAPEUTIC PHARMACOGNOSY

In this book the medicinal plants and plant-derived drugs are classified in two ways:

a) Botanically: for each plant species the family, genus and species, to which it belongs, is given.
b) Pharmacologically: for each plant species and its corresponding drug the pharmacological effects and/or its medicinal (therapeutic) use (is given).

The Anatomical Therapeutic Chemical (ATC) classification that is used here is recommended by WHO and is used in several European countries.

The main groups of the ATC classification system are listed below.
ATC system main groups:

A Alimentary tract and metabolism
B Blood and blood forming organs
C Cardiovascular system
D Dermatologicals
G Genito-urinary system and sex hormones
H Systemic hormonal preparations, excl. sex hormones
J General antiinfectives for systemic use
L Antineoplastic and immunomodulating agents
M Musculo-skeletal system
N Nervous system
P Antiparasitic products
R Respiratory system
S Sensory organs
V Various

In the Anatomical Therapeutic Chemical (ATC) classification system, the drugs are divided into different groups according to their site of action and therapeutic and chemical characteristics.

The drugs are divided into 14 main anatomical groups (1st level) with two therapeutic subgroups (2nd and 3rd level). The 4th level is a chemical/therapeutic subgroup and a chemical substance subgroup is level 5. Each main group is thus divided into 5 levels in the ATC system.

Drugs are classified according to their main therapeutic use, on the basic principle of only one ATC code for each drug.

A drug may be used for two or more equally important indications and the main therapeutic use of a drug may differ from one country to another. Cross-references will be given to indicate the various uses of such drugs.

Drugs classified at the same ATC 4th level cannot always be considered pharmacotherapeutically equivalent, since the adverse drug reaction profile, mode of action and therapeutic effect may differ.

Drugs not clearly belonging to any existing ATC 4th level group of related substances will as a main rule be placed in an x group ('other group').

The fact that a plant-derived drug is mentioned in the following does not necessarily mean that it is recommended by the authors.

A Alimentary tract and metabolism

A01 Stomatological preparations

The rinsing of one's mouth and the process of gargling is used both prophylactically to prevent the establishment of gingivitis and therapeutically to alleviate the symptoms of inflammation in the oral cavity. Many pharmacopoeias include *Gargarisma* as the Latin term for gargling solutions based on natural products such as essential oils, tannins and mucilages.

Essential Oils Generally, essential oils exert astringent and bacteriostatic effects. Three drugs are mentioned here:

PEPPERMINT OIL, *Menthae piperitae aetheroleum* This oil is produced by steam distillation of the flowering parts of fresh peppermint. This plant is a hybrid between three different *Mentha*-species (*M. silvestris L.*, x *M. rotundifolia* x (L.) Huds. x *M. aquatica* L.), family *Lamiaceae*, which contains appr. 1% essential oil. Economically peppermint is the most important of all plant drugs with an annual production of more than 1,000 tons. The essential oil contains about 50–70% menthol, partly free and partly esterified. Three different qualities are on the market: a) pure menthol, b) the complete oil containing menthol and c) peppermint oil, where menthol is taken away, thus a more or less menthol-free oil.

Menthol has spasmolytic, choleretic, antiseptic, local anaesthetic and antiinflammatory effects.

The oil is used in gargling-water.

SAGE, *Salviae folium* is the leaf of *Salvia officinalis* L., family *Lamiaceae*. The leaves contain an essential oil with the main constituents thujone, cineole, camphor, borneol. The essential oil has an antiseptic and fungicidal effect, which is used in gargling-water and in some toothpastes.

CLOVES, *Syzygii aromatici flos* are the flower buds of *Syzygium aromaticum* (L) Merr. et Perry, family *Myrtaceae*, which is cultivated in several tropical countries. It contains 12–26% essential oil with 70–95% eugenol.

Both the essential oil and pure eugenol exert antiseptic and analgesic effects and are used for the treatment of local infections in tooth-cavities and for killing the tooth nerves.

Tannins The use of tannins for gargling solutions is based on their astringent and antiinflammatory effects:

RHATTANY ROOT, *Rathanhiae radix* is the root of *Krameria triandra* Ruiz et Pav, family *Krameriaceae*, a semibush, which grows in the Andes (Peru). The root is mostly used as a tincture usually together with tincture of myrrh for the local treatment of inflammations in the oral cavity.

TORMENTIL, *Tormentillae rhizoma* is the rhizome of *Potentilla erecta* (L), Räusch, family *Rosaceae*, a plant belonging to the European flora. It is used in the same way as Rhattany root and it has also been used internally as an antidiarrheal drug.

Mucilage-containing drugs The polysaccharide mucilage prevents physical and chemical irritation of the epithelium of the mouth cavity and pharynx by forming a protective layer over the mucous membrane.

LANCE-LEAF PLANTAIN, *Plantaginis lanceolatae herba* is the aerial part of *Plantago lanceolata*, family *Plantaginaceae*, and contains mucilage of the arabinogalactose type and tannins. The drug is used as a mild astringent mucilage for treatment of catarrh in the respiratory and alimentary tracts.

MALLOW, *Malvae folium and malvae flos* is prepared from *Altheae officinalis.*, L., family *Malvaceae*. A mucilage consisting of D-galacturonic acid, D-galactose, L-arabinose and L-rhamnose occurs in all parts of the plant. The content of mucilage is highest at the beginning of the flowering period.

The mechanism of action is the same as described for other mucilages.

A02 Antacids, drugs for treatment of peptic ulcer and flatulence

A02A Antacids

A product, deglycyrrhizinised liquorice, in which the mineralocorticoidal effect is eliminated: the exact mechanism of action is not known, but it provokes an increased production of mucin, increased number of new cells in the gastric mucosa and a spasmolytic effect. It is frequently formulated with some inorganic antacids.

AO2B X Antiulcer Agents

LIQUORICE Liquiritae radix is the stolon and root of *Glycyrrhiza glabra* L., family *Fabaceae*. The major constituents of liquorice include glycyrrhizin which is the calcium and potassium salts of glycyrrhizinic acid which in turn is the diglucuronide of a triterpene aglycone called glycyrrhetic acid. Liquorice also contains the isoflavone liquiritin and polysaccharides. The plant, its extracts and β-glycyrretic acid have antiinflammatory and mineralocorticoid activity of relevance to the treatment of peptic ulcers, because of the structural similarity of β-glycyrretic acid to cortisone. However, the mineralo-corticoid activity can result in serious side-effects due to the pseudoaldosteronism related to the corticosteroid effects. These include oedema, sodium retention and hypokalaemia leading to hypertension. The flavonoids also show antispasmolytic and antiulcer activity. Glycyrrhizinic acid has been semisynthetically modified by replacing the diglucuronide with succinic acid to give the antiulcer drug carbenoxolone. It is also marketed as an effective treatment for aphthous or mouth ulcer.

A02D Antiflatulents

As a pharmacological group, the *carminatives* are the drugs that diminish flatulence. Plants containing essential oils have been used for many centuries. Generally these drugs, owing to their taste and flavour, stimulate directly or by reflex mechanisms gastric secretion and intestinal motility and thus also resorption. It has been shown that the part of the stomach into which the oesophagus opens is relaxed by essential oils.

Furthermore, some essential oils have pronounced spasmolytic effects. On tests with guinea-pig ileum, melissa oil, peppermint oil, clove oil and angelica oil exerted the strongest spasmolytic effect, and among the pure ingredients of essential oils tested, it was found that eugenol, caryophyllenoxide, citral and citronellal had the strongest effect. No spasmolytic effect was shown by the essential oils of *Carvi fructus, Anisi fructus* and *Foeniculi fructus.*

Therefore it has to be anticipated that the demonstrated carminative effect of these essential oils does not depend on the spasmolytic but on the stimulatory effect on intestinal motility.

In the evaluation of the spasmolytic efficacy of preparations containing the essential oils with carminative action one has to distinguish between a mixture of essential oils and alcoholic total extracts of the drugs.

Alcoholic extracts of *Carvi fructus, Foeniculi fructus* and *Matricariae flos (chamomillae flos)* show a clear spasmolytic effect (but not the essential oils!) which depends on other ingredients, like flavones.

The following drugs containing essential oil are used for their carminative action: *Menthae piperitae folium* and *aetheroleum, Foeniculi fructus* and *aetheroleum, Carvi fructus* and *aetheroleum, Coriandri fructus* and *aetheroleum, Angelicae radix* and *aetheroleum, Cardamoni fructus* and also three bitter drugs containing pungent substances: *Calami rhizoma, Galangae rhizoma* and *Zingiberis rhizoma*. The three last-mentioned drugs are dealt with under section A09B.

PEPPERMINT, *Menthae piperitae aetheroluem* The drug is used as a stomatological preparation (A01) and has spasmolytic, cholagogue and antiseptic effects besides its carminative action.

FENNEL, *Foeniculi fructus* and *aetheroleum* The drug exerts a carminative effect and is also dealt with in section RO5C, Expectorants.

CARAWAY, *Carvi fructus* and *aetheroleum* Caraway is the schizocarp of *Carum carvi* L., family *Apiaceae*, growing in Asia and Europe, now also grown in the USA and Canada. The drug contains 3–6% essential oil containing 50–80% (+) carvone and limonene. It has spasmolytic, antibacterial and carminative effects.

ANISE, *Anisi fructus* and *aetheroleum* Anise is the schizocarp of *Pimpinella anisum L.*, family *Apiaceae*, with its origin in Egypt and Greece and cultivated in Europe and South America. The drug contains 2% essential oil with the main ingredient anethole (80–90%). The oil is also used as an expectorant (section R05C).

CORIANDER, *Coriandri fructus* and *aetheroleum* Coriander is the schizocarp of *Coriandrum sativum* L., family *Apiaceae*, growing in the Eastern Mediterranian region and now cultivated in Southern Europe, Georgia and India. The globe-shaped fruit contains essential oil with the principal ingredients D-linalool (60–70%) and α-pinene, as well as α- and γ-terpinene.

Spasmolytic and carminative effects are present.

EUROPEAN ANGELICA, *Angelicae radix* The drug is the root of *Angelica archangelica* L., family *Apiaceae*, which grows commonly in the mountain regions of Scandinavia and Northern Asia. It contains 0.3–1.0% essential oil, with α- and β-phellandrene (20–40%), α-pinene (up to 30%) and bitter substances: furocoumarin α-derivate angelicin and coumarin derivatives xanthotoxin, imperatorin and umbelliferone.

The root has spasmolytic, cholagogue and carminative effects. Sunbathing should be avoided when used over long periods, since the furocoumarins make the skin sensitive to UV-light.

A02E Antiregurgitants

This group comprises preparations with effect on gastro-oesophageal reflux. Alginic acid and aluminium hyperoxide in combination are used.

ALGINIC ACID Alginic acid is a polysaccharide that occurs in various brown seaweeds. It is extracted from the seaweed as soluble sodium salt and the acid is regenerated by precipitation using dilute H_2SO_4, calcium chloride or ethanol. Alginic acid and its insoluble salts can be spun into fibres for use as surgical dressings. Because it is highly viscous it provides a gel which prevents food and gastric juices (acid and enzymes) regurgitating up the oesophagus, thus preventing symptoms such as heartburn etc.

A03 Antispasmodic and anticholinergic agents and propulsives

Spastic contractions in smooth muscles in the gastrointestinal, biliary and urinary tracts and uterus may be caused by parasympathetic tonus, gallstones or kidneystones or ischaemia. It is of great practical clinical interest to be able to relieve these spasms, whatever their aetiology.

The classic experimental procedure used to investigate spasmolytic effect is to provoke a spasm in an isolated piece of the intestine of rabbit. The capacity to inhibit a spasm provoked by acetylcholine is called a neurotropic effect (given by atropine), whereas a direct relaxing effect within muscle cells is called musculotropic (given by papaverine).

A03A Spasmolytics of papaverine-type

Papaverine is one of the alkaloids in *Opium*, which is described in NO2A. Papaverine levels range from 0.8–1.5% in Opium, from which it can be extracted, but is also synthetised to a large extent. Papaverine lacks any CNS-effect, but inhibits spasms in smooth muscles with a point of action in the muscle cell itself. It is generally used in combination with analgesics and is administered subcutaneously or orally.

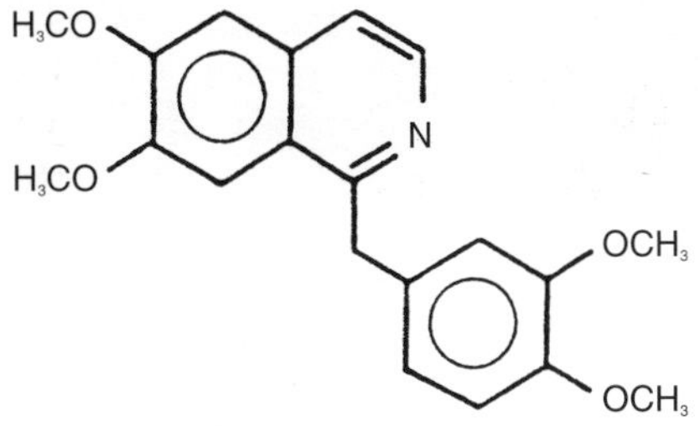

Papaverine

Spasmolytic activity with papaverine-like mode of action may be found in:

SILVERWEED (Goosegrass), *Anserinae herba.*

CORIANDER, *Coriandri fructus* (cf. AO2D)

SAGE, *Salviae folium* (cf. AO1)

YARROW, *Millefolii flos.*

CHAMOMILE, *Matricariae flos (Chamomillae flos)*

Chamomile tea produced from the flowers of *Matricaria recutita* has long been used as an antispasmodic for gastric and menstrual cramps. Studies with guinea-pig ileum have shown that the water-soluble flavonoid fraction and especially apigenin has antispasmodic activity three times that of papaverine.

ROMAN CHAMOMILE, *Anthemidis flos, Chamomillae romanae flos* (cf. NO2B)

AO3B Belladonna and derivatives

Naturally occurring tropane alkaloids and semisynthetic derivatives are used as spasmolytics because of their anticholinergic (-parasympatholytic) effect through competitive inhibition of acetylcholine at receptors in smooth muscles.

These alkaloids are from a structural point of view, esters between the alcohols such as tropine in e.g. hyoscyamine and scopine in e.g. scopolamine and different aliphatic and aromatic acids, mainly tropic acid. The tropane alkaloids within the family *Solanaceae* belong mainly to the tropine series, whereas the Coca-alkaloids from the family *Erythroxylaceae* mainly belong to the pseudo-tropine series.

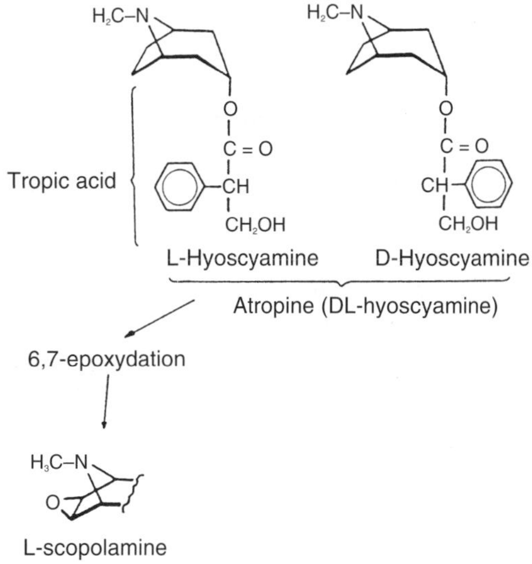

Tropic acid

L-Hyoscyamine D-Hyoscyamine

Atropine (DL-hyoscyamine)

6,7-epoxydation

L-scopolamine

The clinically used Solanaceae-alkaloids hyoscyamine, atropine and scopolamine are esters of tropic acid. Since tropic acid has an asymmetric C-atom, these alkaloids occur in L-, D- and DL-forms, the L-form of which is the also most pharmacologically active.

During the extraction procedure a partial racemisation takes place. The racemate of L-hyoscyamine is atropine. The pharmacologically active isomer is L-hyoscyamine, which has double the activity of the racemic atropine. L-hyoscyamine is used for the inhibition of gastric secretion in cases of gastric and duodenal ulcer.

Atropine is the most important and most used anticholinergic drug. Besides its use as spasmolytic agent and inhibitor of secretion, atropine is used as an antidote in cases of overdosage of cholineesterase inhibitors (physostigmine and organo phospate insecticides) and for dilation of the pupil in cases of iritis (inflammation of the iris). Atropine has a central stimulant action, which is most obvious after overdosage. Hallucinations might occur, which is the explanation of the effect, which was noted during the days of the witchcraft trials, after overdose of henbane.

Drugs that contain tropane alkaloids include the following:

BELLADONNA LEAF, *Belladonnae folium* (Fig. 2) is the dried leaf of *Atropa belladonna* L., family *Solanaceae*, which is a perennial herbaceous plant, about one metre high, growing in Central Europe, but nowadays cultivated in former Jugoslavia, Hungary, Poland, China and India. The content of total alkaloids should be 0.35% according to the European Pharmacopoeia: 75% of this content is L-hyoscyamine, the rest is mainly atropine, which is formed during the drying process. Both extract and isolated pure substances are used clinically.

STRAMONIUM LEAF, *Stramonii folium* is the dried leaf of *Datura stramonium L.*, family *Solanaceae*, with violet flowers and spiny capsule. Originally from Central America it is now spread globally in warm regions. The alkaloid content is 0.2–0.6%, with L-hyoscyamine and L-scopolamine found in the ratio 2:1 in mature plants.

Fig. 2. *Atropa belladonna* L.

HENBANE LEAF, *Hyoscyami folium* is the dried leaf of *Hyoscyamus niger L.*, family *Solanaceae*, originally growing in Europe, nowadays spread all over the globe. The alkaloid content is 0.04–0.17%, with L-hyoscyamine and L-scopolamine found usually in the ratio 1.2:1 in mature plants.

For the extraction of L-hyoscyamine, Egyptian henbane, *Hyoscyamus muticus*, is used, owing to its high alkaloid content 0.5–1.5%.

The Australian species *Duboisia myoporoides R.Br.*, family *Solanaceae*, is a bush-tree, which grows on the east coast of Australia, New Caledonia and adjacent islands. Its leaves contain about 2% alkaloids. Leaves harvested in October contain hyoscyamine as the main alkaloid, whereas leaves harvested in April have scopolamine as the main alkaloid. (Fig. 3)

Fig. 3. *Duboisia myoporoides* R.Br.

AO4 Antiemetics and antinauseants

AO4AD Other antiemetics

SCOPOLAMINE Scopolamine has peripherial anticholinergic effects, but its effect on the central nervous system (motor and vegetative centres) is depressive in contrast to atropine. Therefore scopolamine is used for premedication before operations (together with morphine) and as a remedy against seasickness.

Scopolamine is extracted from the leaves of *Duboisia myoporoides, R.Br.* and also from aerial parts of *Datura sanguinea* Ruiz et Pav., cultivated in Equador. (Fig. 4)

Semisynthetic alkaloids are also used, e.g. methylhomatropine and methylscopolamine where the balance between the anticholinergic and the CNS-effects have been changed. Drowiness and visual disturbances could occur with such products.

GINGER ROOT, *Zingiberis rhizoma* The drug is the rhizome of *Zingiber officinale* Roscoe, family *Zingiberaceae.* Its origin is South East Asia, but it is now cultivated on a large scale in India, several countries in Africa and the West Indies, and on a small scale in several tropical countries. The drug contains an essential oil with 60% zingiberene and zingiberol (with a pronounced smell) and a phenolic mixture of gingerols and related shogaols, with a pungent taste. This drug is mainly used as a spice.

Fig. 4. *Datura sanguinea* Ruiz et Pav. Cultivation in Equador

Ginger stimulates the heat receptors in the stomach, which gives a burning feeling of heat. It also dilates the peripheral blood vessels and increases perspiration, which gives a cooling of the skin. Further, it has a carminative and spasmolytic effects on the gastrointestinal tract. Zingerone, which is formed from gingerol, exerts an antiinflammatory effect, used for both acute and chronic inflammations, and its effectiveness in rheumatoid conditions is under test.

There have been 4 animal studies (in frogs, suncus, dogs and rats) which have shown positive antiemetic activity when the animals were dosed with extracts of ginger or some of the isolated gingerols. In those studies vomiting was induced with chemotherapeutic agents such as cis-platin and cyclophosphamide. In human studies, the results have been mixed. Eight of the 14 studies of ginger as an antiemetic have shown positve antiemetic effects. Four studies showed that ginger was as effective as standard synthetic antitravel sickness products. Indeed some studies showed that ginger was superior. Ginger has also shown benefits in cases of vomiting due to hyperketonaemia in children, postoperatively after gynaecological surgery and in cases of *hyperemesis gravidarum* (morning sickness). Three studies have shown no benefits in travel sickness, two studies in postoperative gynaecological cases were also negative and one study in patients on cancer chemotherapy was inconclusive. It is suggested that the effects of ginger are on the gastrointestinal system and hence one of its major advantages is the lack of the drowsiness associated with other antiemetics. It is known that constitutents of ginger are mutagenic and this might be a cause for concern if the plant were to be used by pregnant women for morning sickness. However, while there have been no scientific studies of potential teratogenicity, the fact that there are also antimutagenic compounds present, allied to the widespread and lengthy use of ginger as a spice and in confectionery, would seem to suggest a low potential for harm.

AO5 Bile and liver therapy

AO5A Bile therapy

AO5A A Bile acid preparation

OX BILE, *Fel tauris* Ox bile is obtained from cattle in slaughterhouses, and contains sodium salts of some 20 conjugated bile acids. Among these, cholic acid and desoxy-cholic acid occur in the ratio 10:1.

A preparation containing cholic acid and dehydrocholic acid has a pronounced choleretic effect, which is mainly due to dehydrocholic acid, and which compared to cholic acid has a less pronounced effect on the digestion of fats and their resorption.

A05A X other drugs for bile therapy
Two categories within this group have long been recognised. Thus, choleretics are substances which increase the outflow of already formed bile from the liver and gallbladder, while cholagogues or cholekinetics stimulate the production of bile in the liver. Most drugs mentioned here have both effects.

The therapeutic indication for this category is inflammation in the bile tract, no outflow of the bile and formation of gallstones.

Medicinal plants belonging to this group are often mixed in clinical use. The most important species are the following:

CELANDINE, *Chelidonii Herba* is the aerial part of *Chelidonium majus*, L., family *Papaveraceae.* It has mainly a choleretic effect. It contains about 20 alkaloids (0.4–0.8%) with the main alkaloid chelidonine, which also has a spasmolytic effect (about half the effect of that of papaverine).

BOLDO LEAVES, *Boldo folium* are the leaves of *Peumus boldus* Molina, growing in the lowlands of central Chile (family *Monimiaceae*). The main alkaloid is boldine (ca. 0.25%), which has choleretic, spasmolytic effects, and increases gastric secretion and has also sympatholytic activity.

BARBERRY ROOT, *Berberidis radix* is the root of *Berberis vulgaris*, L., family *Berberidaceae*, containing 3% alkaloids with berberine, jateorhizine and palmitine as the main constitutents. It stimulates the smooth muscles and empties the gallbladder but has no choleretic effect.

TURMERIC, *Curcumae rhizome* is the rhizome of *Curcuma longa*, L., (syn. *Curcuma zanthorrhiza* Roxb.), family *Zingiberaceae.* It contains curcumin (dicinnamoylmethane derivatives) and essential oil.

Experimentally a cholekinetic and antiphlogistic effect has been demonstrated.

DANDELION ROOT, *Taraxaci radix* is the root of *Taraxacum officinale*, L., family *Asteraceae*. The drug contains bitter substances like lactucopicrine (taraxacine), taraxasterol and phytosterols. Experimentally the choleretic effect in dogs and rats has been demonstrated and as a bitter drug it stimulates gastric secretion.

AO5B Liver therapy, lipotropics

Since there is no causal therapy of the liver, only symptomatic therapy has been used and is used.

The results of treatment with plant-derived liver therapeutics are improvement or normalisation of the biochemical data (seen after a blood analysis) and disappearence of various secondary symptoms. In patients with liver cirrhosis (alcoholics) the survival rate is improved.

MILK THISTLE *Silybi marianae fructus*, which is the fruit of *Silibum marianum* (L.) Gaert, family *Asteraceae*. The active ingredient in the fruit is silymarin (a mixture of isomeric flavonolignans with silibinin (silybin) as the most active substance, effective as an antidote to poisoning with fly agaric (phalloidin and α-amanitin).

Silymarin has the ability to increase the regeneration of liver cells by stimulation of protein synthesis. Besides the liver-protecting action it also has choleretic, cholekinetic and spasmolytic effects.

ARTICHOKE, *Cynarae folium* is the leaves from *Cynara scolymus* L., family *Asteraceae*. The extract of artichoke leaves has a liver-protecting effect, increases the regeneration of liver cells and has choleretic and lipid-lowering effects.

Extracts of the fruit of the Chinese plant *Schisandra chinensis* (Turcz) Baill., family *Magnolinceae*, have been shown to have a clinical effect on hepatitis B virus.

AO6 Laxatives

Man has since times immemorial used many products from the plant kingdom or minerals to stimulate and facilitate defecation. The procedure of evacuation has from time to time been surrounded by religious ideas, which have influenced alternative names for laxatives such as cathartics, purgatives and aperients (words of Greek and Latin origin, meaning to purify or clean).

Laxatives are especially useful for persons who cannot endure the exertion of defecation, especially patients with certain types of hernia or severe hypertension. Most laxatives are used to counteract acute constipation.

Laxatives are mainly grouped according to their mode of action. Enemas are classified in one group A06A G, regardless of mode of action.

There are several modes of action of laxatives. One group acts as softeners, which can be administered orally or as an enema. Another group is the bulk producers; these are taken orally and swell in the intestine, and thus stimulate the defecation reflex. A large group are the stimulant laxatives with an action in the colon by inhibiting the absorption of electrolytes and water through a specific pharmacological mechanism.

AO6A B Contact laxatives Two different types of laxatives, which have a primary effect on intestinal motility, are the anthranoid glycosides and castor oil.

Anthranoid laxative plants The anthranoid-containing plants are essentially similar in their mode of action. In this group the active substances are a group of polynuclear or tricyclic hydrocarbon derivatives based on anthracene. Various oxidised and reduced forms constitute the aglycones of a variety of 'C' and 'O' glycosides. The fully oxidised form is the anthraquinone form, with the partially reduced form represented by the anthrone nucleus. Dimeric forms whereby two anthracene nuclei are chemically linked to form dianthrones are especially important. All of the medicinally important anthranoids are 1,8 dihydroxy derivatives, a feature which appears to be essential for activity. Different aglycones arise through substitution at carbons 3 and 6 on the benzene rings which are characteristic of these compounds. The plants containing these compounds have been used as laxatives for many years. They include:

SENNA FRUIT (SENNA POD), *Sennae fructus* The drug is the dried fruit of either *Cassia senna* L., known as Alexandrian Senna, or of *Cassia angustifolia* Vahl, known as Tinnevelly Senna, family *Fabaceae*. (Fig. 5) Alexandria Senna comes from the Sudan and Somalia, whereas the Tinnevelly species is cultivated in Southern India. Both plants contain a complex mixture of anthranoids, chiefly Rhein-Rhein dianthrone diglucosides known as Sennosides A+B (see formula) as well as small amounts of other dianthrone diglucosides, monoanthraquinone glucosides and aglycones. There is, however, a distinct difference in the amount of hydroxyanthracene compounds in each type of Senna, which is why Pharmacopoeias distinguish between them. Alexandrian Senna contains not less than 3.4% of hydroxyanthracenes which for analytical purposes are calculated as total Sennoside B. Tinnevelly Senna, on the other hand, contains not less than 2.2% of hydroxyanthracenes.

Fig. 5. *Cassia augustifolia* Vahl. Cultivation in Tinnevelly, South India

SENNA LEAF, *Sennae folium* The dried leaflets of both of the Senna species are also used medicinally. The leaves of both contain the same amount (not less than 2.5%) of anthracenes and for this reason Pharmacopoeias permit the use of either species or of a mixture of the leaves of both the Alexandrian and Tinnevelly types.

FRANGULA BARK, *Frangula cortex* The drug consists of dried bark from stems and branches of *Rhamnus frangula* L., family *Rhamnaceae*, which grows in Europe and Western Asia. Because the fresh bark contains anthracene glucosides based on the anthrone and dianthrone glycoside forms, which have a harsher laxative effect, the bark is stored for at least 1 year or heated to allow the conversion of the reduced anthrones to oxidised forms. This results in the formation of a mixture of glucofrangulins A+B (emodin diglycosides) (see formula) and their partial decomposition

Glucose–O O OH

L-Rhamnose–O CH_3

O

Glucofrangulin A

products, frangulin A+B (emodin monoglycosides), as well as other anthracenes in smaller amounts. Frangula bark of pharmacopoeial quality must contain not less than 6.0% of glucofrangulins calculated as glucofrangulin A.

CASCARA, *Rhamni purshiani cortex* Cascara bark is more widely used than Frangula as a laxative in the USA since the trees of *Rhamnus purshianus* DC, family *Rhamnaceae*, are cultivated in Northwestern USA and Canada. Cascara has also been known as Cascara Sagrada from the Spanish name for sacred bark. Like the closely related Frangula bark, fresh Cascara bark contains reduced emodin-based glycosides. In order to convert these to milder oxidised forms the bark must either be stored for 1 year or heat treated.

Cascara bark should contain not less than 8.0% of hydroxyanthracenes. Not less than 60% of this mixture must be in the form of cascarosides which are complex mixed 'O' and 'C' glycosides based on the anthrone nucleus. Loss of the 'O'-glucose gives rise to secondary glycosides notably Aloin A+B, which also are found in Aloes.

ALOES Many different types of Aloes have been used in medicine as laxatives but the two main types now recognised are Cape Aloes from Southern Africa and Barbados Aloes from the Caribbean. Cape Aloes consists of the concentrated and dried juice of the leaves of various species of Aloe, mainly *Aloe ferox* Miller and its hybrids. Barbados Aloes is produced from *Aloe barbadensis* Miller (also known as *Aloe vera* L.), family *Liliaceae*. (Fig. 6) The different Aloe plants are all succulent cactus-like plants with large fleshy water-storing leaves. These are cut near the base and the juice from the pericyclic cells is collected. This juice is then concentrated by heat and allowed to dry into dark masses.

Medicinal Aloes contains C-glycosides similar to those of Cascara as well as primary glycosides called Aloinosides and also various resins. The main constituent is

Fig. 6. *Aloe barbadense* Mill (A. vera L.) flowering plant

Barbaloin. Pharmacopoeial-quality Cape Aloes must contain not less than 18.0% of anthracene derivatives.

The European Pharmacopoeia also include *Aloes extractum siccum normatum*, which is a standardised dry extract made by treatment with water and addition of sugar. It contains 19–21% hydroxyanthracene derivatives.

In a dose of 0.1 g Aloes is a laxative and in a dose of 0.2–0.5 g it is a purgative, i.e. a laxative with a very strong laxative effect.

Aloe gel is the colourless, mucilaginous gel obtained from the parenchyma cells in the central region of the leaves of *Aloe barbadensis* Mill. It has a dermatological use – see DO3 C.

RHUBARB ROOT, *Rhei radix* Medicinal Rhubarb consists of the peeled and dried root of *Rheum* species, chiefly *Rheum officinale* Baill. and *R. palmatum* L., family *Polygonaceae*, but not of *Rheum rhaponticum* L., the common garden rhubarb. Rhubarb, like Ginseng and Ephedra, is one of the oldest and most widely used Chinese drugs. Chemically it contains a most complex mixture of anthracene glycosides, the most important of which are the dimeric Sennosides. Tannins are also found which can counteract the laxative effect and at low doses (0.05–0.5 g) it is their astringent effect that predominates. Higher doses of 1–2 g have a mild laxative effect. Extracts are also used in the treatment of mouth ulcers.

Medicinal uses of anthranoid laxatives
All of the 1,8 dihydroxy anthracene-containing plants have basically the some mode of action. The glucosides tend to act as pro-drugs preventing the absorption of the molecules in the stomach and small intestine. When the glucosides reach the colon they are metabolised by enzymes from the bacterial flora of the large intestine into the active anthrone forms. For example, in the case of Senna drugs the sennosides are converted into rhein-anthrone. Cascarosides are known to be hydrolysed into the corresponding aloins which are subsequently reduced to the aloe-emodin and chrysophanol anthrones. The glucofrangulins from Frangula are converted into emodin-anthrone.

There are two different mechanisms of action for the various anthrones which are seen as the main active metabolites with specific activity on the colon. One mechanism involves the anthrone influencing the motility of the colon through an inactivation of the enzyme Na+/K+ATP-ase which results in an inhibition of the Na+ pump and of C1-channels in the colonic membrane. This results in the speeding up of transit time through the bowel. A second mechanism involves enhanced fluid secretion into the lumen of the bowel by stimulation of mucous and chloride. The resulting net transfer of fluid into the intestinal lumen triggers reflex defecation. There is a delay of 6–12 hours from administration of the laxative until defecation occurs due to the time taken for the glycosides to reach the colon and be transformed into the active anthrones.

It is recommended that stimulant laxatives should not be used for more than 2 weeks because long-term use may cause inflammation of the mucous membranes and pigmentation of the colon (*pseudomelanosis coli*) which is reversible after use is stopped. There is the risk of the well-known laxative abuse syndrome, particularly in anorexia nervosa and in bulimia. This leads to diarrhoea, disturbance of electrolyte and water balance and an atonic colon. The hypokalaemia which develops can result in cardiac and neuromuscular problems and there may also be albuminuria and haematuria.

While there are no reports of undesirable or harmful effects during pregnancy, the usual precautionary measures should be taken during the first three months of pregnancy, particularly for preparations likely to contain free emodin and aloe-emodin which have given positive results in tests for mutagenicity. These results have also caused concern in terms of possible carcinogenicity but studies in animals and case control studies in humans do not show any increased risk of bowel cancer from usage of anthranoid laxative at recommended doses.

Non-anthranoid contact laxatives

CASTOR OIL, *Ricini oleum* This is the fixed oil pressed from the ripe seeds of *Ricinus communis L.*, family *Euphorbiaceae.* This oil contains 75–80% of triglycerides of the fatty acid ricinoleic acid (12-hydro-9,10-cis-octadecenoic acid). Intestinal lipases can release this acid which has local stimulating effects on motility similar to that of the anthranoids.

Castor oil seeds contain a toxic glycoprotein called Ricin which remains in the pressed seed cake but which can be removed with water.

MOUNTAIN ASH, *Sorbus aucuparia* L., Family *Rosaceae* has been used as a mild laxative in the form of the parasorbinic acid-containing fresh juice of the berries. Conversely the high content of pectin and tannins in the fruits means that the boiled berries can be used to treat diarrhoea.

AO6A C Bulk Producers Within this group there are fibres, which consist of plant polysaccharides and lignins. Fibres can be divided into non-soluble and gel-forming. The non-soluble fibres are resistant against any effect of the intestinal bacteria, while the gel-forming fibres are degraded by the bacteria of the colon.

Pectin, guar and glucomannans are examples of gel-forming fibres, whereas cellulose, lignins and seed coats are examples of non-soluble fibres. The non-soluble fibres bind water and swell and keep water in the intestine, while the gel-forming fibres are degraded, which make them unsuitable as laxatives.

Some kinds of plant mucilages and gums have the ability to substantially increase in volume on contact with water, particularly in the alkaline milieu in the intestine where they can increase their volume by 40 times. In this way they exert a mechanical effect on the intestinal walls in the colon and rectum, which facilitates the passage of faeces.

ETULOSE Etulose is a cellulose ether, which is not absorbed in the intestine, but swells substantially in the stomach and forms a homogenous gel, which retains water and gives the content of the intestine a loose consistency and bigger volume. Etulose should be mixed with food and water so that the content of the intestine is homogenous.

STERCULIA GUM, *Sterculiae gummi* (Karaya Gum) is obtained from *Sterculia urens* Roxb. and related species within the family *Sterculiaceae*. The gum oozes out of natural cracks or after incisions in the stem of *Sterculia* species, grown in India.

The gum that has exuded solidifies to semitransparent, colourless droplike structures, which have a weak smell of acetic acid. The sterculia gum is insoluble in water, but swells considerably: 1 g gum can bind more than 40 g water, and thus decreases water reabsorption from the colon, and increases in volume. This volume - increasing effect (bulk effect) contributes to an increased intestinal motility. In the treatment of patients with chronic constipation sterculia gum is preferred to stimulant laxatives.

ISPAGHULA SEED, *Psyllii semen* The drug is obtained from *Plantago ovata*, Forskål, family *Plantaginaceae*, grown in the province of Gujarat, India. Ispaghula seeds are lighter in colour than the South European seed from the species *Plantago afra* L. (= P. *psyllium* L.) and *Plantago indica* L. In commerce it is the isolated seed coats (testa) consisting of epidermis with an adjacent layer that usually occur. Like sterculia gum, Ispaghula works on the principle of a hydrophilic colloid effect. The volume-increasing property of Ispaghula affects the peristalic reflex and contributes to the establishment of physiological motility.

LINSEED (Flax), *Lini semen* is harvested from *Linum usitatissimum* L., family *Linaceae*. The plant has been cultivated for centuries in Europe for textile purposes. In the epidermis of the seed coat there is 3–6% mucilage consisting of galacturonic acid, galactose, rhamnose, and arabinose, mannuronic acid. Linseed is cheap and simple to use as a laxative: 1 spoon of linseed is allowed to swell overnight in a cup of water, and next morning the contents of the cup are swallowed. Its value has been demonstrated in two studies of patients with constipation.

GUAR-BEAN, *Cyamopsidis semen*, Guar-bean is obtained from the endosperm of the seed of *Cyamopsis tetragonolobus* (L) Talbert, family *Fabaceae*. Guar powder contains up to 88% water-soluble mucilage, which can be extracted and fractionated by precipitation with ethanol. The main part is a galactomannan with a mol. weight of ca 22,000, made up of D-mannose and galactose-molecules in the ratio 2:1. Guar is used as a laxative, but it also decreases the level of blood sugar (A10B) and cholesterol (LDL).

It is imperative that patients taking bulk laxatives be advised to maintain fluid intake because of the risk of physical obstruction of the intestine or oesophagus by what is termed a phytobezoar. It is also noteworthy that the absorption of other drugs taken at the same time as bulk laxatives can be delayed. This particularly affects coumarin anticoagulants, digoxin, antibiotics, paracetamol and also oral contraceptives. Since many bulk laxatives are formulated into slimming products, this is an area where pharmaceutical care and advice are essential.

AO6AD osmotically acting laxatives
These drugs contain inorganic and organic salts, which are very poorly absorbed, binding water in the intestine, resulting in stools of light consistency.

Fruit acids (such as malic acid in plums, tartaric acid in tamarind, or malic, tartaric, citric acid, in figs) have analogously a laxative effect.

AO7 Antidiarrheals, intestinal antiinflammatory/antiinfective agents

AO7B Intestinal absorbents

A07BA Charcoal preparations

CARBO ACTIVATUS *Ph Eur.* The drug is used in the treatment of acute oral poisonings. It efficiently binds various remedies and toxins and prevents the absorption of orally administered substances.

AO7C Electrolytes with carbohydrates

Salt solution for rehydration has, according to WHO, the following formula: Glucose 20 g; sodium chloride 3.5 g; sodium carbonate 2.5 g; potassium chloride 1.5 g and water to 1,000 ml.

AO7D Antipropulsives

This group comprises agents which reduce gastrointestinal motility.

OPIUM TINCTURE, *Opii tinctura* The tincture is an effective and cheap antipropulsive to treat the enteritis that European travellers encounter in tropical and subtropical regions. Dose: 10 drops of tincture (10%) in a glass of water 1–2 times per day. Besides its antipropulsive effect, the opium tincture reduces the stomach-ache, given by the intestinal spasms, which is often the most painful symptom. Generally it can be stated that the most important thing in cases of acute enteritits is to substitute the fluid that is lost with various salt solutions containing potassium and sodium ions. (see AO7C).

CALUMBA ROOT, *Calumbae radix* Other antipropulsive drugs include Calumba root which is the root of *Jatrorrhiza palmata*, Miers, family *Menispermaceae*, a wild-growing liana in East Africa. It contains the alkaloids jatrorrhizine and palmatine with effects on smooth muscles like morphine.

CAROB FRUIT, *Ceratoniae fructus* The drug is the pod of *Ceratonia siliqua L.*, family *Cesalpiniaceae*, a tree cultivated in the Mediterranean zone. The pods have several seeds, containing sugar, proteins, pectin and mucilage. The fresh fruit has a sweet taste and is eaten. During the Second World War it was observed in Spain that starving children who had been eating carob had less stomach trouble than children who had not. Therefore, a 10% decoction of dried carob was introduced against dysentery, enteritis and dyspepsia. The therapeutic effect is attributed to pectin, mucilage and fruit acids.

CAROB SEED *Ceratoniae semen* Carob contains 40% water-soluble mucilage in the endosperm. The seeds are boiled with 4% sodium carbonate solution and then washed. The coat is separated, the seeds are pressed and the hard endosperm separated from the other ingredients. The endosperm contains up to 90% of the polysaccharide carubin, which upon grinding gives a powder, which is used as such or after water extraction and evaporation gives purified carubin.

Carubin is a polysaccharide (galactomannan with mol. w. ca. 310,000), which after hydrolysis gives 16–20% D-galactose and 80–84% D-mannose. It is used therapeutically for the treatment of vomiting in small children, where the mucilage gives a thickening of the stomach contents. The tannins have a more or less specific effect to prevent the fixation of coli-enterotoxin into the cells of mucous membrane of the intestine.

BILBERRY, *Myrtilli fructus* Bilberry is the fruit of *Vaccinium myrtillus, L.*, family *Ericaceae*, and contains 5–10% tannins of the catechin-type, anthocyanins, flavone glycosides and organic acids. The tannins in bilberry and in other tannin-containing drugs precipitate proteins in the epithelium of the intestine. In this way the absorption of toxic substances and secretion is diminished. This is the mode of action in the treatment of diarrhoea.

The anthocyanins from bilberry are also used for the adaptation of the eyes to darkness (see SO1X).

Several different types of active substances are found in the group of drugs for the treatment of diarrhoea. For Example: pectins in apple, mucilage in marshmallow and tannins in tormentil. Peppermint and *Asafoetida* are considered to be effective in cases of irritable bowel syndrome.

A07F antidiarrhoeal microorganisms

On the market there are preparations containing mixtures of dried bacteria used in diarrhoea and for normalisation of the bacterial flora (after oral antibiotic therapy). An example is the following mixture: *Streptococcus faecium, Streptococcus thermophilus, Lactobacillus acidophilus, and Lactobacillus bulgaricus.*

AO8 Antiobesity preparations, exclusive diet products

AO8A A Centrally acting antiobesity products Centrally acting compounds mainly used to produce anorexia, e.g. phentermine, are classified in this group.

AO8A B Peripherally Acting Antiobesity Products Drugs to support slimming are:

KONJAC ROOT (Konnyaku) *Amorphophalli radix* is the root of *Amorphophallis konjac*, Kikock, family *Araceae*, cultivated in Japan and China. It contains polysaccharides, mainly glucomannan, also arabinose and galactose.

Glucomannan binds (like other mucilages) large volumes of water, and thus fills the stomach and gives a more satisfied feeling without giving too many calories. Care is needed regarding fluid intake and interactions with other drugs.

GUAR BEAN, *Cyamopsidis semen* The drug contains galactomannan with properties similar to Konjac root.

AO9 Digestives, incl. enzymes

AO9A A Enzyme Preparations Enzymes can be used therapeutically. Some are of animal origin (from the stomach and pancreas of food animals), some of vegetable origin and others of microbiological origin. Examples of the first type include pepsin and pancreatin, a mixture of amylase, lipase and trypsin (a protease) which catalyses the metabolism of proteins into smaller peptides and amino-acids. Pancreatin is frequently prescribed for patients with cystic fibrosis.

Aspergillus species especially *A. oryzae*, are the source of cellulases and hemicellulases which help break down vegetable foods and allow normal digestive enzymes access to the nutrients in plant cells. In addition, enzymatic metabolism of the normally indigestible cellulose of plant cell walls reduces excessive formation of intestinal gas.

PINEAPPLE, *Ananas comosus* (L). Merr., Family *Bromeliaceae* and
PAPAYA, *Carica papaya*, L., Family *Caricaceae* are two plants which contain proteolytic enzymes: bromelain in pineapple and papain in unripe papaya fruits. These two are proteases and can degrade protein in meat in cases where there is a deficiency of digestive enzymes. Papain is widely used as a meat tenderiser as are ficin from *Ficus glabra* and the bromelains. It is also used in more purified form (Chymopapain) in chemonucleolysis to shrink ruptured or slipped vertebral discs.

AO9B Bitter Drugs

The use of *tonics*, strengthening drugs and *amara* (bitter drugs) aims to increase subjective well being and also strengthen the appetite. Bitters usually do not contain the pure substances, but water and alcoholic extracts of plants containing substances with a bitter taste and other ingredients which are pharmacologically relatively inactive.

Substances with a bitter taste do not constitute a chemically distinct group, and for that reason the bitter drugs are not standardised by means of a chemical method, but with an organoleptic method, i.e. determination of the bitterness value. A test panel tries to find the lowest concentration, that still has a bitter taste. An example: if a bitter taste is felt with a solution of 1 g Gentian dissolved in 30,000 ml water, but not at a higher dilution, its bitterness value is 30,000.

The medicinal plants mentioned here stimulate the appetite and the secretion of gastric juice and bile, and therapeutically they should be administered $\frac{1}{2}$ hour before a meal.

The bitter drugs also contain other substances, which is the basis for their classification into five groups.

AO9B A Amara Aromatica Besides the bitter substances, essential oils are present which give an aromatic, spice-like taste.

ABSINTHE, *Absinthii Herba* The drug is the dried aerial part of *Artemisia absinthium*, L., family *Asteraceae.* The plant has a strong bitter and aromatic taste. It grows in Asia, Europe and Northern Africa. The bitter taste is given by absinthin (sesquiterpenelactone 0.15–0.4%). The bitter value of the drug is 10,000–25,000). It also contains 0.5% essential oil containing chamazulene, thujone, thujol, and phellandrene; its content and composition depend upon climate, habitat and time of collection.

Absinthe has been used for the fabrication of absinthe liqueur in Central Europe, but is nowadays forbidden owing to the toxicity of thujone to the CNS leading to headache, dizziness, cramps and delirium. Now there exist thujone-free varieties of the plant with even stronger stimulant effects than the 'normal' absinthe.

EUROPEAN ANGELICA, *Angelicae Radix* The drug is the root of *Angelica archangelica* L., family *Apiaceae.* The root contains 0.3–1.0% essential oil with phellandrene (70–80%), the bitter substance angelicin (furocoumarin derivative) and osthol and osthenol (coumarin derivatives). Phellandrene has a central stimulating effect, which later on can be inhibitory. At normal dosages the root is mainly an aromatic bitter drug which stimulates the appetite, though the plant can cause allergic reactions.

YARROW, (Milfoil) *Millefolii flos* Yarrow is the flower of *Achillea millefolium* L., family *Asteraceae.* The flowers contain up to 0.5% essential oil. One part of the essential oil is transformed to chamazulene (up to 40%).

The bitter substances include the sesquiterpene lactones millefolide and acetylbalchanolide. The essential oil contains 8–10% cineole, which has an antiseptic and expectorant effect; chamazulene is antiinflammatory and stimulates the granulation process. The bitter substances stimulate the digestion and also have a mild spasmolytic effect.

BITTER ORANGE, *Aurantii pericarpium* is the dried fruit peel of the bitter orange, *Citrus aurantium* L. ssp. *amara* (L.) Engl., family *Rutaceae.* All citrus-fruits have a peel, which consists of two layers: the exterior yellow-orange-coloured *flavedo*, and an interior soft, white layer, called the *albedo.* In this drug most of the albedo should be removed. This drug is an aromatic bitter containing essential oil, with terpenes such as(+)-limonene, geranial and its isomers. The bitter substances belong to two structural types, the flavone glycosides naringin and neohesperidin, their bitter taste depending on the structure of the sugar moiety. To the other group belongs the terpene-derivative limonene. Therapeutically the drug is used to stimulate the appetite and gastric secretion. It is used as a spasmolytic and sedative.

AO9B B Amara-acria containing bitter and pungent substances

GINGER, *Zingiberis rhizoma* is described under AO4, Antiemetics.

GALANGA, *Galangae rhizoma* is the rhizome of *Alpinia officinarum* Hance, family *Zingiberaceae*, cultivated in South-East Asia. It contains pungent substances, including various diaryl heptanoids, gingerol and flavonoids (bitter substances). The effect of the drug is spasmolytic, antiphlogistic and antibacterial. It is used to treat anorexia and dyspepsia, and stomachache.

AO9B C Amara-adstringentia containing bitter substances and tannins

CONDURANGO, *Condurango cortex* Condurango is the bark of *Marsdenia condurango*, Reichenb., family *Asclepiadaceae*, a tree growing in the mountain regions of South America. It contains 1–2% of condurangin, a glycoside with a bitter steroid aglycone. The bitterness value is approximately 15,000. The bark also contains tannins.

CINCHONA BARK, *cinchona cortex* is described under section PO1 B.

AO9B D Amara-mucilaginosa containing bitter substances and mucilage

ICELAND MOSS, *Lichen islandicus* consists of the dried thallus of *Cetraria islandica (L.)* Acharius sensu latiore, family *Parmeliaceae.* Characteristic constituents include bitter-tasting lichen acids: usnic acid gives the bitter-taste and has antibacterial effects. It also contains polysaccharides, which form a mucilage: Lichenin, a linear cellulose-like polymer of D-glucose, is soluble in hot water, and on cooling it forms a gel. The other polysaccharide is isolichenin, a linear starch-like polymer of D-glucose, soluble in cold water.

In the form of decoctions Iceland moss is used to treat dry coughs. Here the antibiotic and bacteriostatic lichen acids may be of significance. A further indication is lack of appetite and gastro-enteritis.

BLESSED THISTLE, *Cardui benedicti herba* is the dried, aerial part of *Cnicus benedictus* L., family *Asteraceae.* It is an annual thistle-like herb with sharp thorns on the leaves. The bitterness value is about 1,500, due to the presence of 0.2% cnicin and benedictin. It also contains mucilages, which makes the drug suitable for treatment of catarrh in the upper respiratory tract and the whole gastrointestinal tract.

AO9B E Amara-pura containing only bitter substances

GENTIAN ROOT, *Gentianae radix* is the dried fermented underground organs of *Gentiana lutea* L., family *Gentianaceae.* (Fig. 7)

Fig. 7. *Gentiana lutea* L. Cultivation in South Germany

It contains secoiridoid bitter glycosides, including gentiopicroside, and the extremely bitter amarogentin (its bitterness, value is 1.58 million, i.e. 300 times more bitter than quinine), as well as oligosaccharides, including bitter-tasting gentianose and gentiobioside. The sugar gentianose is formed during fermentation in the drying of the root. This process gives a darker colour and a more pronounced aroma. The bitterness value of the drug is 10-30,000. Amarogentin and amaroswerin, which occur in other *Gentiana* species, are the most bitter substances known up to now. The therapeutic indications for gentian root are anorexia, e.g. after illness, and dyspepsia.

BUCKBEAN, *Menyanthidis folium* is the leaf of *Menyanthes trifoliata* L., family *Gentianaceae*, a water plant distributed on the Northern hemisphere. The leaf is three-

Fig. 8. *Quassia annara* L. Cultivation in Indonesia

foliate and its bitterness value is 4-10,000 due to the secoirodoid glycosides foliamenthin and menthafolin and its biogenetic precursor loganin.

QUASSIA, *Quassiae lignum* L. is the wood from *Quassia amara* L., or *Picrasma excelsa* Planchon, family *Simarubaceae*. (Fig. 8)

The bitterness value is about 4 million, due to quassin and neoquassin. These two bitter substances are the therapeutically active molecules, and their individual bitterness value is about 17 million.

The drug is used to stimulate the appetite and facilitate digestion.

A10 Drugs used in diabetes

A10A Insulins

Insulin is used in the treatment of Diabetes mellitus type1. Insulin was obtained as a by-product of the meat industry. The sequence of the amino acids in human and pig insulins differs only in *one* amino acid (B10), which in human insulin is threonine and in pig insulin alanine.

Since the mid-1980s human insulin, manufactured through a recombinant DNA technique, has become available for clinical use. With this technique insulin is manufactured by producing the A- and B-chains in separate bacterial cultures of *Escherichia coli*

K12; then the two chains are combined to give the insulin molecule.

A10B Oral blood glucose lowering drugs

Increased blood sugar in diabetes type II can be treated with diet or glucokinines.

In the plant kingdom there are plants that have an 'insulin-like' effect. They are called glucokinines.

Such glucokinines are found in *Galega officinalis*, family *Fabaceae*, which contain Galegin 0.5% in the seeds. This plant is nowadays used in traditional medicine contains, but the synthetic biguanide derivative (metformin) is used clinically. An extract of the fruits of the Indian tree *Jambul (Syzygium cumini = Eugenia jambolana)* is used clinically on a limited scale.

Guar gum decreases the level of blood sugar (AO6A C).

A 11 Vitamins

The vitamins, enzymes and hormones belong to the group of biocatalysts, substances which in small quantities must be given to humans and animals for normal function and growth.

The vitamins cannot be synthesised in the organism; in certain cases the organism is capable of carrying out the last step from provitamin to vitamin (β-carotene to retinol). Some vitamins function as redox-catalysts in the metabolism (vitamins A,C,E and K), whereas others function as co-enzyme or activators of enzymes (vitamin B1, B2, B6, B12, folic acid, biotin and vitamin D).

Antivitamins (vitamin antagonists) are substances that specifically decrease or inhibit the action of a vitamin (e.g. Actinomycin is a vitamin D antagonist). Vitamins are not remedies in the usual sense, because vitamin supply with food is a necessary condition for the normal functioning of the metabolism of the organism.

Vitamins B and C are water soluble, whereas vitamins A, D, E and K are fat soluble.

The main indication for vitamin therapy is a threatened or established status of deficiency, which is rather rare in Western Europe and the developed world.

In a few cases vitamins in very high doses exert a pharmacological effect, e.g. vitamin A, nicotinic acid and vitamin D.

In other cases a pharmacological effect has been postulated, but not been proven, e.g. vitamin C.

Any generally tonic and beneficial effect of supernormal doses of vitamins has not been proven.

Most vitamins are non-toxic in very high doses with the exception of vitamins A and D, which can give severe poisoning. Other vitamins can give pronounced discomfort. Multivitamin preparations are usually harmless.

VITAMIN A (*Retinol, Vitaminum A*) Occurs mainly in cod-liver oil and halibut-liver oil, *Jecoris aselli oleum* and *Jecoris hippoglossi oleum*, respectively, and in small amounts in milk, butter, carrots and egg yolk. In earlier times vitamin A was extracted from cod-liver oil, nowadays it is synthesised (=nature-identical vitamin A).

PROVITAMIN A (*B-carotene*) Occurs in carrot, tomato, and paprika. One molecule of β-carotene is transformed in liver and intestinal mucosa to two molecules of vitamin A (retinol). Lack of vitamin A injures the sight and gives xerophthalmia (turbidity) of the conjunctiva, hemeralopia (nightblindness) and increased sensitivity to infections.

VITAMIN B_1 (*Thiamine, aneurine*) Vitamin B_1 has an important function in carbohydrate-metabolism as a prosthetic group in the enzyme co-carboxylase.

Vitamin B_1 occurs in raw rice, coarse bread, milk products, potato, yeast, ground nuts, and peas. Nowadays synthetic vitamin B1 is also used.

VITAMIN B_2 (*Riboflavine, lactoflavine*) Riboflavine plays an important role in the 'respiration-chain'. It occurs in milk products, egg, yeast, tea and coffee. It is manufactured synthetically or micro-biologically by aerobic cultivation of *Ashbya gossypii* and *Eremothecium ashbyii.*

VITAMIN B6 (*Pyridoxine*) As pyridoxal-5-phosphate, Vitamin B6 functions as a co-enzyme in amino acid decarboxylases and transaminases. It occurs in meat (liver), fish (salmon), carrot, avocado and other fruits, egg yolk and yeast. Synthetic material is also used.

VITAMIN B_{12}: SEE BO3 B

VITAMIN C (*L-ascorbic acid*) Vitamin C plays an important role as a redox-catalyst in metabolism in teeth, bone, cartilage and parenchyma-tissue. Vitamin C occurs in several fruits such as an rose hips (*Cynosbati fructus*), Black currant, Sallow-thorn (*Hippophae rhamnoides*), Citrus-fruits, Paprika, Potato, Cabbage, and Parsley.

Ascorbic acid is manufactured synthetically from glucose via sorbitol.

Rose hip, *Cynosbati fructus*, is the dried flower bottom of *Rosa canina* L., family *Rosaceae*, deprived of the hairy fruits ('kernels') from the inside of the drug. The drug contains at least 1% ascorbic acid, and also carotenoids and flavonoids with vitamin P-effect. Rose hips are still used as a vitamin C source in vitamin preparations.

VITAMIN D Vitamin D_2 = *ergocalciferol (calciferol)* Vitamin D_3 = *cholecalciferol*

Vitamin D stimulates the resorption of calcium and organic phosphate from the calcium deposits in the skeleton.

In vitamin D-deficiency in children, rachitis appears with deformities owing to decalcification and disturbances in growth. Therefore vitamin D is given prophylactically to children and pregnant women. Vitamin D-deficiency in adults is unusual in Western Europe. It may occur in vegetarians, especially when combined with absence of sunlight and dairy products.

Vitamin D_2 (ergosterol) occurs in yeast, wheatgerm oil and ergot, while vitamin D_3 occurs in cod-liver oil, egg yolk and butter.

Vitamin D_2 is manufactured by UV-radiation of ergosterol and vitamin D_3 is obtained from tuna fish liver oil and UV-radiation of 7-dehydrocholesterol.

Cod-liver oil, *Jecoris aselli oleum*, is manufactured from the liver of the cod, *Gadus morrhua*, or other species of *Gadus*. The oil is melted from the liver into containers filled with carbon dioxide. Then the oil is chilled to –5°C, and the solid particles are precipitated and separated. The content of vitamin A and D is determined and standardised.

VITAMIN E (*Tocopherol*) Vitamin E comprises tocopherols with similar physiological effects α, β, γ, δ, tocopherol, of which α-tocopherol has the strongest biological effect on foetal development in rats and on their normal growth, but its importance for humans is not known and the existence of deficency symptoms are also uncertain.

In spite of this, therapeutic use in certain forms of internal bleedings, neuritis, skin diseases and threatened abortion has given promising results.

Vitamin E protects polyunsaturated fatty acids from oxidative degradation at the double bonds. It is probable that an excessive intake of polyunsaturated fatty acids low in tocopherol, e.g. corn oil and cottonseed oil, is a relative indication for E-vitamin administration. The tocopherol might diminish the tendency to form 'free' radicals, which are supposed to have pathogenic effects. Tocopherols occur in wheatgerm oil and also in leaves of vegetables, egg yolk and meat. Tocopherol is also synthesised for therapeutic purposes.

No definite effects have been obtained from treatment with large doses of vitamin E to improve fertility and potency.

VITAMIN K: see BO2

A13 Tonics

This group comprises preparations used as tonics etc., if preparations do not fill the requirements to be classified as iron preparations, vitamin preparations etc.

Tonics are non-specific acting drugs, which can be used when a specific therapy is not known or recommended. Since several conditions without clear diagnosis are included here, tonics are frequently used in self-medication.

One indication for tonics is dystonia (diminished tonics in various organs) where tonics will restore the function.

NUX VOMICA *Strychni semen* The drug is the seed of *Strychnos nux-vomica* L., family *Loganiaceae* a tree from South-East Asia (India, Sri Lanka) where it is cultivated. (Fig. 9)

The ripe fruit is similar in shape, colour and size to a small orange and has jelly-like, white, bitter fruit flesh, with 4-5 flat grey seeds. The active ingredients in the seeds are the indole-alkaloids strychnine and brucine; both have a very bitter taste.

Strychnine (used as water-soluble nitrate) diminishes, by antagonism of the inhibitory transmitter substances, the synaptic resistance in the spinal cord and thus increases the tonus of the skeleton (skeletal) muscles. This effect is achieved by subconvulsive doses of 0.1-0.2 mg. Tired, asthenic patients thus get a subjective stimulating feeling of greater physical efforts. This is certainly of therapeutic value - and strychnine is therefore used as an ingredient (*in low dose*) in tonics.

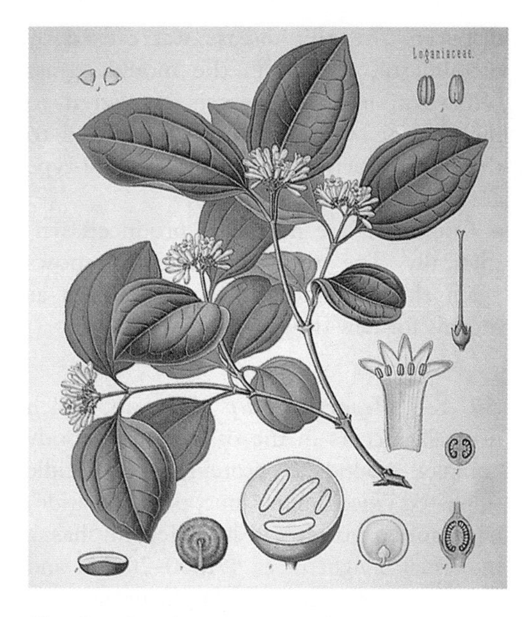

Fig. 9. *Strychnos nux vomica* L.. Flowering branch.

In higher doses strychnine gives tonic convulsions, which could be lethal due to respiratory paralysis, thus giving a toxic and not a therapeutic effect.

Brucine has a weaker convulsive effect (about 1/50). Since this alkaloid has a very bitter taste, it is used as a reference compound for the determination of the bitter value of drugs.

It is also used to denature alcohol in after-shave solutions and similar cosmetics.

B Blood and bloodforming organs

B01 Antithrombotic agents

B01A Antithrombotic agents

B01A A Vitamin K antagonist This group comprises vitamin K antagonists such as the dicoumarol group. The plant *Melilotus officinalis* (L.) Pallas, family *Fabaceae*, that has become mouldy due to microorganisms, caused disease in grazing cows, 'Sweet Clover Disease', where the cows died from internal bleeding. The discovery that this disease was caused by dicoumarol, present in the mouldy grass (coumarin in fresh grass is converted to dicoumarol by the microorganism), led to the introduction of dicoumarol-type anticoagulants, e.g. Warfarin.

Coumarins are nowadays produced synthetically. The story of dicoumarol shows again the usefulness of natural drugs as models for new drugs.

B01A B Heparin group Heparin, which normally occurs in the tissues of the body complex bound to protein, is an acidic sulphated glucosamine (mucopolysaccharide) with anticoagulant effects. Heparin has a molecular weight of ca 17,000–20,000 and is built up of alternating D-glucuronic acid and D-glucosamine residues with 1.4-β-binding. Heparin is obtained from the intestinal mucosa from pigs and then purified.

Heparin inhibits blood coagulation both *in vivo* and *in vitro.* In combination with a co-factor, antithrombin III, heparin interferes with several steps in the coagulation process. Furthermore, heparin activates lipoprotein-lipase, which splits triglycerides into free fatty acids and glycerol.

Heparin is administered intravenously or subcutaneously. The maximal anticoagulant effect is obtained within a few minutes after intravenous injection, 2–3 hours after slow intravenous infusion and within 2–4 hours after subcutaneous injection. The indications are the prophylaxis and treatment of thromboses and embolism of the lungs.

B01A D Enzymes Streptokinase For medical purposes a highly purified streptokinase preparation is used. It is produced by cultivation of a *β-haemolytic Streptococcus*-strain, belonging to Lancefield group C. After several purification steps the culture solution is purified by removal of various other enzymes such as streptodornase, streptolysin and hyaluronidase.

Streptokinase is a protein with a molecular weight of about 50,000. It is colourless and easily soluble in water. The substance occurs on the market in freeze-dried ampoules mixed with buffer-salts. Streptokinase has the specific effect of transforming plasminogen in blood to the fibrinolytically active enzyme plasmin. The activity is measured in International Units (IU).

Plasmin is a proteolytic enzyme with specific affinity for fibrin of clots. The indications for streptokinase are: acute cardiac infarction, venous and arterial thrombosis, lung and arterial emboli.

B02 Antihaemorragics

B02A Antifibrinolytics

B02B Vitamin K and other haemostatics

B02B A Vitamin K (Phyllokinon) Vitamin K_1 = *phytomenadione*, Vitamin K_2 = *pharnokinon.*

The side chain in the natural vitamins is not necessary for the biological effect. Synthetic 2-methyl-1.4-naphthoquinone has a good vitamin K effect and is available on the market as a coagulation vitamin.

Natural vitamin K occurs in spinach, tomatoes and vegetable oils. Vitamin K_1 can be extracted from spinach, and Vitamin K_2 from rotten fishmeal. Vitamin K_1 and K_2 are formed in the intestine through the activity of coliform bacteria.

In the case of Vitamin K-deficency owing to bile obstruction, which has lowered the value for prothrombin complex, vitamin K1 is administered intravenously. In moderate vitamin K-deficiency induced by therapy with antibiotics or diarrhoeas of long duration, oral vitamin K therapy is needed. Prophylactically vitamin K is used for treatment of haemorrhage in the newborn and therapeutically in cases of overdosage of warfarin-type anticoagulants.

B03 Antianaemic preparations

B03B Vitamin B12 and folic acid

B03B A Vitamin B_{12} (Cyanocobalamin) The naturally occurring cobalamin has attached to the Cobalt-atom a desoxyadenyl residue, which is substituted by a cyanide group (by addition of KCN-solution), thus giving the modified natural product cyanocobalamin. Hydroxycobalamin occurs in the liver and can be produced by the effect of light on a slightly acidic solution of cyanocobalamin.

For therapeutic use vitamin B_{12} is no longer extracted from the liver of food animals, but is obtained from submersion cultivation of either *Streptomyces griseus, Streptomyces olivaceus, Bacillus megatereum*, or *Propiobacterium shermanii.*

B03B B Folic acid and derivatives Folic acid is transformed by ascorbic acid to leucovorin, which (like vitamin B_{12}) is necessary for some transmethylation processes, among others for the synthesis of desoxyribonucleic acid and ribonucleic acid. Deficiency in folic acid causes megaloblastic anaemia of the same type as deficiency in Vitamin B_{12}.

Therapeutically folic acid is used in megaloblastanaemia, sprue, and after therapy with hydantoin derivatives and therapy of leukaemia with folic acid antagonists.

Folic acid is produced synthetically.

B04 Serum lipid reducing agents

FIBRES

POLYUNSATURED FATTTY ACIDS

GAMMALINOLENIC ACID (see Chapter VII)

BO4A X

GARLIC BULB, *Allii sativi bulbus* Garlic bulb consists of fresh or dried compound bulbs of *Allium sativum* L., family *Alliaceae.* This is a very old spice and medicinal plant, which is of great importance in self-medication in Europe.

The European pharmacopoeia defines Garlic powder as produced from the bulbs of *Allium sativum* L. cut and freeze-dried or dried at a temperature not exceeding 65° and containing not less than 0.45% of Allicin. Allicin does not occur in fresh garlic or in carefully dried material. It is produced by the action of an enzyme called alliinase on the main sulphur-containing amino-acid called alliin ((+)-S-allyl-L-cysteine sulphoxide). Other characteristic, genuine constituents are (+)-S-methyl-L-cysteine-L-sulphoxide, gamma-glutamyl peptides, and ubiquitous amino-acids.

In the presence of the enzyme allinase, alliin will be converted to allicin:

$$CH_2=CH-CH_2-\underset{\downarrow \atop O}{S}-CH_2-\underset{| \atop NH_2}{CH}-COOH \xrightarrow[\text{Allinase}]{H_2O} CH_2=CH-CH_2-\overset{O \atop \uparrow}{S}-S-CH_2-CH=CH_2$$

Alliin Allicin

In turn, allicin is the precursor of various transformation products, including ajoenes, vinyldithines, oligosulphides and polysulphides, depending on the conditions applied.

Material derived from garlic by steam distillation or extraction in an oily medium contains various allicin transformation products.

The essential oil has a strong smell and the sulphur-containing constituents give the characteristic smell and taste. Depending on the method of production the amount of active compounds in the oil is highly variable, which has an influence on the pharmacological effects.

As mentioned above, allicin is not stable and in the presence of water and the oxygen in the air it is transformed to polysulphides which give the strong characteristic smell of the air expired from the lungs - the same reaction takes place in the human body as in the test tube.

The therapeutic indications for garlic bulb are prophylaxis of atherosclerosis; treatment of elevated blood lipid levels, insufficiently influenced by diet; improvement of the circulation in peripheral arterial vascular disease; and respiratory infections and catarrhal conditions.

Garlic bulb also has a choleretic and spasmolytic effect, which explains its use as a carminativum, whereas the disinfectious action is the basis for its use in the case of dyspepsia and chronic intestinal infections. Furthermore, an aqueous extract of fresh garlic bulb has bactericidal and fungicidal effects. Also, effects on the immune system have been reported.

B05 Plasma substitutes and perfusion solutions

B05A Blood and related products

B05A A plasma substitutes and plasma-protein fractions

DEXTRAN The dextrans are mucilaginous polysaccharides, which are built up from D-glucose molecules.

The dextrans are obtained microbiologically from sucrose by fermentation with microbes from the genera *Leuconostoc, Acetobacter* and *Streptococcus*, which contain the enzyme dextransucrase.

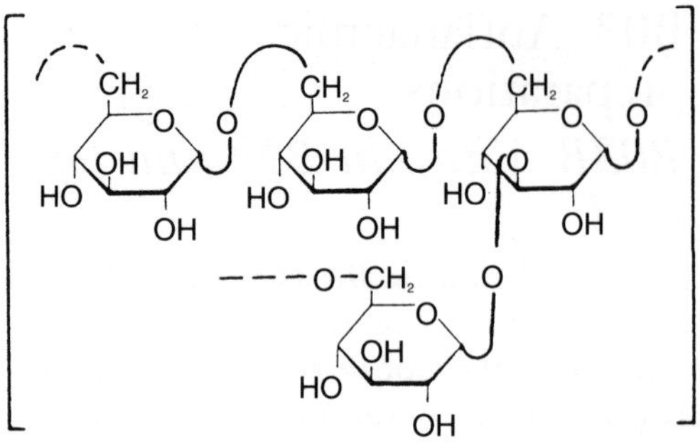

Dextran

For the production of dextran for medical use the Swedish producer Pharmacia-Upjohn uses the species *Leuconostoc mesenteroides* B512, which gives a dextran with low content of side chains, which is preferable for medically used dextran.

The crude dextran (with great variation in molecular weight from a few thousand up to several millions) undergoes partial acidic hydrolysis, which is followed by fractionation to two main products with different molecular weights.

The biological properties of dextran are a function of its molecular structure, molecular weight and distribution of molecular weights. Dextran 70 gives an increase in plasma volume, corresponding to the volume injected. After infusion of Dextran 70, about 30% is excreted by normal renal function within 6 hours and in all about 40% within 24 hours. The rest is degraded in the body at a rate of 70 mg/kg bodyweight per day.

Dextran is used clinically as a colloid osmotic plasma expander in case of shock and prophylactically in cases of postoperative and post-traumatic tromboembolic conditions.

GELATIN, *Gelantina* Gelatin is a degradation product of collagen with a molecular weight of 60,000–90,000 which upon hydrolysis gives 95% amino-acids. Gelatin is obtained from the skins and bones of animals slaughtered for food. Gelatin swells in cold water, dissolves in warm water and solidifies on cooling.

Gelatin polymerisate is produced by adding a solution of aliphatic di-isocyanate to gelatin, hydrolised to a molecular weight of 12,000. In this reaction a branched product with molecular weight of 30,000 and also a globular one is formed. It is the active principle of a preparation used in cases of shock of different origins and degrees of seriousness. About 85% of the amount injected is excreted unchanged via the kidneys, 10% via faeces and 3% is metabolised. Half-life is about 5 hours.

ALBUMIN Albumin is a normal plasma protein with two main functions: to maintain the colloid osmotic pressure in blood plasma and bind and transport low molecular substances: e.g. bilirubin, fatty acids, hormones and certain drugs. Administration of 1 g albumin increases the circulating plasma volume by about 18 ml. At normal serum albumin levels, 40–50 g/l albumin is responsible for about 80% of the colloid osmotic pressure.

Albumin is produced from human plasma by the alcohol fractionation method according to Cohn. The albumin is then controlled for the absence of antibodies against HIV, HCV and hepatitis B surface antigen (HBsAg). Furthermore, the albumin solution is kept at 60° C for 10 hours to inactivate HIV and hepatitis viruses.

The indications are hypovolaemic shock when other therapy is insufficient, and also pronounced hypoalbuminaemia.

B05B Intravenous solutions

B05B A Solutions for parenteral nutrition

FAT EMULSION For parenteral nutrition a fat emulsion for intravenous administration is used. Preparations contain a fractionated soya-oil emulsified with some fractionated egg phospholipids. About 60% of the fatty acids are essential fatty acids: the particle size and biological properties are similar to those in natural chylomicrones.

SOYA OIL, *Sojae oleum* Soya oil is obtained from seeds of *Glycine max* (L.) Merr., family *Fabaceae*. The soya bean as cultivated in the USA, South America and India, is up to 1 m high, a bushy, very hairy annual plant with small blue-violet flowers in racemes. The fruit is a 1 cm hairy pod, which contains 2–5 round, pea-like seeds. The seeds contain about 20% oil, which is produced commercially by pressing. The oil contains glycerides of linoleic acid (50%), oleic acid (30%), linolenic acid (7%) and saturated fatty acids (13%).

Carbohydrates

GLUCOSE, *Glucosum* Glucose occurs freely in sweet fruits, in grapes and honey together with fructose. Commercially glucose is obtained by hydrolysis of starch from potato or corn. After hydrolysis, performed with diluted hydrochloric acid under pressure, the mixture is neutralised, filtered, decolourised with charcoal, concentrated and crystallised. Purification is done by recrystallisation. Glucose 55 mg/ml is isotonic with the blood.

Glucose is also available as a thick syrup (*glucosum liquidum*). Invertose (= invert sugar) is a mixture of equal parts of glucose and fructose, which is obtained by hydrolysis of sugar (sucrose). Fructose is more rapidly metabolised than glucose. For that reason an invertose solution 100 mg/ml can be injected with the same speed as a glucose solution 55 mg/ml without any substantial loss of sugar in the urine. Invertose-solution is a suitable base solution for the addition of electrolytes.

AMINO ACIDS On the market there are preparations containing 18 essential and non-essential amino-acids which are required for synthesis of the proteins of the body. The balance between the amino-acids is such that a positive nitrogen balance is obtained both in nutrition over long time periods and in the postoperative phase. In order that the amino-acids should be used optimally for protein synthesis, the need for energy should be supplied by administration of carbohydrate and fat. Glucose is preferable as a source of carbohydrates for intravenous nutrition.

B05B C Solutions producing osmotic diuresis

MANNITOL Mannitol is the active ingredient in the drug *Manna*, which is the dried juice after incision of the bark of *Fraxinus ornus* L., family *Oleaceae*, a small tree which is cultivated in special Sicilian gardens called 'frassinetti'.

Mannitol is not metabolised and is excreted unchanged via glomeruli with only very small reabsorption in the tubules. Mannitol is therefore used diagnostically for determination of the filtration capacity in the glomeruli. The diuretic effect of mannitol depends upon an osmotic effect in the tubules and increased capillary circulation. Pure mannitol is used in solution for parenteral administration.

The drug *Manna* with 50–60% Mannitol is used as a mild laxative. Mannitol is not absorbed from the intestine and gives increased intestinal bulk because of the absorption of water.

INULIN Inulin is a polymer of D-fructose molecules (fructofuranose), which are linked to each other in β-2.1 bonds. There is a terminal glucose molecule. Inulin occurs as a reserve food instead of starch in plants, mainly within the family *Asteraceae*: for example *Inula helenium* L., *Asteraceae*. In the form of a 10% aqueous solution inulin is used in tests of kidney function (like mannitol). Inulin is not absorbed or metabolised. Since there is no reabsorption in the tubules, the rate of excretion is a measure of the function of the glomerulus (inulin clearance).

B05 C Irrigating solutions

In this group are found the natural products mannitol (B05B C) and sorbitol.

Sorbitol is named after *Sorbus aucuparia* L. mountain ash, family Rosaceae (cf A06A B). Sorbitol is found in the berries, but nowadays it is produced by catalytic reduction of D-glucose. Besides its use in irrigating solutions, it is used as a sweetener for diabetic patients and finally it has a mild laxative effect.

C Cardiovascular system

C01 Cardiac Therapy

C01A Cardiac glycosides

The cardiac glycosides are steroid glycosides which exert a specific effect on the dynamics and rhythm of the failing heart. Characteristic features of the cardiac glycosides are a 5- or 6-membered, unsaturated lactone ring at position C17 and 1–5 sugar molecules at position C3. The stereochemistry of the steroid skeleton is also significant. Depending on the nature of the lactone ring two types of glycosides can be distinguished:

1. Cardiac glycosides of the Cardenolide-type (C-23-derivatives) contain a γ-lactone ring with a single double bond. The Digitalis glycosides belong to this type.
2. Cardiac glycosides of the Bufadienolide type (C-24 derivatives) contain a δ-lactone ring with two double bonds. The Squill glycosides and Toad-extracts (Bufa-species) belong to this group.

There are more than 30 different aglycones of cardiac glycosides, which have been isolated from several genera. The sugar portion, which is bound in the C3 position, is represented by the usual sugars such as D-glucose, D-fructose and L-rhamnose and also by sugars that are not found in other glycosides, such as the 2-desoxy sugars digitoxose and cymarose.

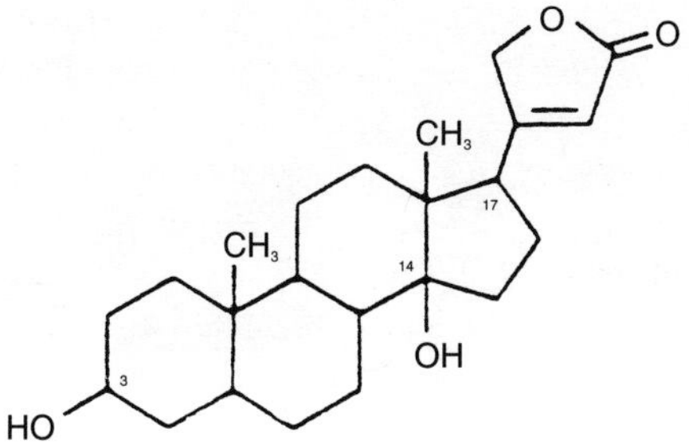

Card -20(22)-enolide

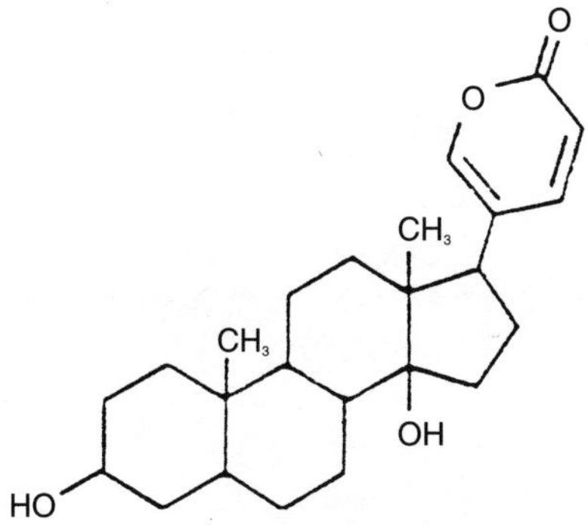

Bufa -20, 22 - dienolide

The glycosidic bonds with the 2-desoxy sugars are more easily hydrolysed by acids than bonds involving glucose. Since there are enzymes in the plant, which most easily split off the terminal glucose molecule, one distinguishes between primary (original) glycosides and secondary glycosides, where the terminal glucose molecule has been removed.

The cardiac glycosides have a bitter taste. They are soluble in alcohol, poorly in water and can be extracted from aqueous solutions with a mixture of chloroform and alcohol, a method that is also used industrially.

The cardiac glycosides have up to now been isolated from a limited number of plant families: *Scrophulariaceae (Digitalis lanata), Apocynaceae (Strophanthus gratus, Nerium oleander), Hyacinthaceae (Urginea maritima), Convallariaceae (Convallaria majalis)*, and *Ranunculaceae (Adonis vernalis)*.

All cardiac glycosides have qualitatively the same effect. Primarily the systolic contraction of the heart muscle is strengthened. The mode of action of the cardiac glycosides, which is not known in all details, depends on rhythmic intracellular liberation of calcium-ions by inhibition of the calcium-ion outflow and an increase in the inflow of calcium-ions into the cell. This takes place by inhibition of Na+/K+-activating membrane-ATP-ase. (*Digitalis receptor*). The concentration of Na+ is increased and that of K+ is decreased intracellularly. In this way the myosin-ATP-ase is activated with improved use of ATP, which gives increased power of contraction by facilitated reaction between actin and myosin.

The toxic effect arises from too strong a decrease in the intracellular concentration of potassium by inhibition of the uptake of potassium in the cell.

It was pointed out previously that the cardiac glycosides have a qualitatively similar effect, but they differ quantitatively as regards absorption, latency, duration of action, and excretion.

Symptoms of poisoning by the cardiac glycosides are loss of weight, discomfort, vomiting (also after parenteral administration), diarrhoea, headache, disturbance of sight and arrythmias. Old people are particularly sensitive. The toxic dose is very variable. The risk for toxic effects is increased in rapid dehydration and hypokalaemia.

C01A A Digitalis Glycosides

FOXGLOVE, *Digitalis purpurae folium*
The drug is the dried leaf of *Digitalis purpurea L.*, family *Scrophulariaceae*:

More than 30 cardiac glycosides of the cardenolide type have been isolated and the total content in the dried leaf (the drug) is 0.30–0.40%. The drug also contains enzymes that split off the sugars from the glycosides. Digitalis is a biannual plant, which upon cultivation can be perennial. During the first year a rosette is developed with hairy leaves.

Whereas the flowering plant with its bell-shaped flowers is very decorative in the garden, only the rosette in the first year is used for drug production, because it is easier to harvest and the content of cardiac glycosides is higher. There are chemical races of *Digitalis purpurea*, the most important being the digitoxin and gitoxin races.

This plant was not used during classical antiquity – it was the pharmacologically similar squill (see pg. 73) that was used. The English physician William Withering introduced foxglove into medicine in 1775 for the treatment of dropsy. His classical work 'An account of the Foxglove and some of its Medical Uses: with Practical Remarks on Dropsy and other Diseases' was published in 1785.

Today one of the *Digitalis purpurea* glycosides is mainly used therapeutically, namely digitoxin, which is the secondary glycoside of Purpureaglycoside A: i.e. where the terminal glucose molecule has been split off.

Digitoxin is absorbed almost completely and has a half-life of 5–7 days. For this reason its concentration is relatively constant. There is only a slight change if the patient forgets a digitoxin dose once in a while; an interval of 1–2 days per week is often recommended to avoid accumulation. Digitoxin is metabolised in the liver and then eliminated.

DIGITALIS LANATA LEAF, *Digitalis lanatae folium* The drug is the dried leaf of *Digitalis lanata* Erh., family *Scrophulariaceae*, which is a hairy perennial or biannual herb. (Fig. 10) In the first year the plant forms a rosette of leaves – which are harvested and used for extraction of cardiac glycosides – and in the second year an aerial stem about 1 m high with grey-brown flowers, which have a redbrown venation and a white lip. The inflorescence is woolly (lanata (Latin) = woolly). This species is indigenous to Central and South-Eastern Europe. It is now cultivated in Holland, Germany, Hungary and the USA.

Fig. 10. *Digitalis lanata* Erh.. Cultivation in the Netherlands

The leaves are sessile, linear-lanceolate to oblong-lanceolate and up to about 30 cm long and 4 cm broad. The margin is entire, the apex acuminate, and the veins leave the midrib (at a very acute angle.) There has been confusion between *D. lanata* leaves and the ribwort plantain *Plantago lanceolata*.

The *Digitalis lanata* leaves from the first-year rosette are nowadays the most important source for extraction of Digitalis glycosides because the recent intensive research on digoxin has made this glycoside the first choice for therapy.

The total content of glycoside in the dried leaf is about 1%. Up to now about 60 glycosides have been isolated from the *D. lanata* leaf. The main glycosides are the two primary glycosides lanatoside A and lanatoside C, at a concentration of about

250 mg/100 g drug, and the other primary glycosides lanatoside B, D and E, which together constitute 100 mg/100 g drug. The industrial extraction of the leaves and the resulting purification steps are concentrated on lanatoside A and C and their hydrolysis products.

Lanatoside C has a rapid onset of action and a rapidly disappearing effect. Digoxin is absorbed less than digitoxin and has a half-life in plasma of about 35–38 hours. Digoxin is easy to control, i.e. the effect appears rapidly and possible side-effects also disappear rapidly. If there are difficulties in oral administration digoxin can be given intravenously. The induction time to maximal effect of digoxin is about 2 hours.

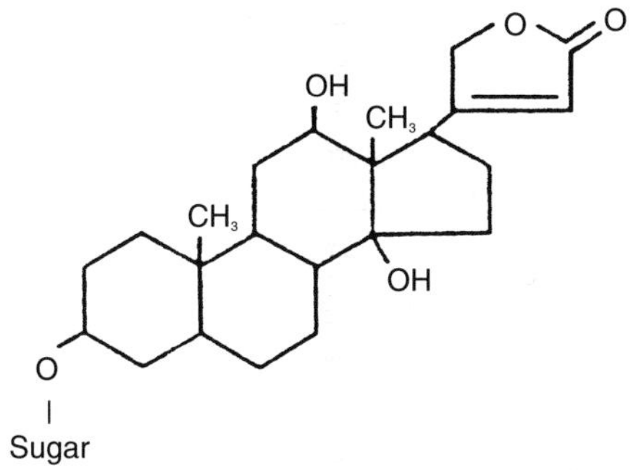

Sugar:- Digitoxose - Digitoxose - Acetyldigitoxose - Glucose
Glycoside: Lanatoside C

C01A B Scilla Glycosides

SQUILL, *Scillae bulbus* The drug consists of the dried sliced bulbs of *Drimia maritima* (L.) Stearn syn. *Urginea maritima* (L.) Bake, family *Hyacinthaceae*, from which the membranous outer scales have been removed (fig. 11).

Among the medicinal plants, that contain cardiac glycosides, squill is the one that has been used earliest in history. By 1500 BC squill was used in Egyptian medicine for the treatment of dropsy (oedema), a symptom of failing function of the heart. The effect of squill was known by Hippocrates and by the physician to Empress Maria Teresia, van Swiesen.

Squill is a perennial herb, growing in the Mediterranian region. Its bulb can reach 30 cm in diameter and 8 kg in weight. This is known in commerce as white squill. English supplies are derived from Malta and Italy where the white variety of squill is grown. Red squill, also derived from a variety of *Urginea maritima*, is collected in Algeria and Cyprus and differs from the white variety in containing red anthocyanin pigment.

Proscillaridin, the aglycone of which has a 6-membered lactone-ring with two double-bonds, is a glycoside of the bufadienolide type and scillaren A constitutes $\frac{2}{3}$ of the total glycoside fraction, whereas the remaining $\frac{1}{3}$ is constituted by some 25 other glycosides. The chemical relation between these two glycosides is the following:

Scillaren A is hydrolysed by enzymes to proscillaridin + glucose, thus, the terminal glucose molecule is split off.

Proscillaridin, which increases the contraction capacity, undergoes a comprehensive first-pass metabolism.

The biological availability is 35–40%. The effect diminishes by 35% per day. The elimination is not influenced by a decreased renal function. Squill is often used as an expectorant because of its ability to cause gastric irritation leading to a reflex secretion from the bronchioles.

RED SQUILL has the same cardiac activity as the white variety. But the most pronounced effect is it's toxicity to rodents. Therefore it has been used as a rat poison and contains in addition to other constituents the glucosides scilliroside and scillirubroside.

C01A C Strophanthus Glycosides

STROPHANTHUS GRATUS SEED *Strophanthi grati semen* is the dried seed of *Strophanthus gratus* (Wall and Hook) *Baill.*, a liana belonging the family *Apocynaceae.*

Fig. 11. *Drimia martima* L. Stearnas, syn., *Urginea martima* L. Baker

From the about 40 *Strophanthus* species, which grow mainly in Equatorial Africa, only two or three species are used in medicine: *S. kombé* in East Africa and *S. gratus* in Equatorial Africa and the Western part of Africa, where also the third species, *S. hispidus*, is grown.

These *Strophanthus* species are lianas with latex and opposite hairy - in *S. gratus* non-hairy - leaves. High up in the heads of the trees, the flowers of the climbing *Strophanthus* lianas are found. The petals are drawn out into ribbon-like flaps, 10–20 cm long, which, however, are missing in the species *S. gratus*. Each flower gives rise to two divergent follicles which, when ripe, are 20–35 cm long and 2–2.5 cm broad and bowed outwards. Each follicle contains 100–200 seeds, which are furnished with feathery awns. These are almost invariably removed from the seed before exportation.

The seed of *S. gratus* is brown; compressed and somewhat twisted; margin sharp-angled, appears glabrous, whereas all the other species are hairy with abundant trichomes.

Strophanthus seeds are still used as an ingredient in arrow poisons. In Eastern Cameroon and adjacent parts of the Central African republic the seed of *S. gratus* is used under the name Nea (French: graine de Nea), whereas the seed of *S. kombé* is used along the Zambesi River for preparation of gombi - or kombe - arrow poison.

A co-traveller of Livingstone, Dr Kirk, discovered by chance the effect of the arrow poison on the heart in 1859. His toothbrush had been in the same bag as the samples of the arrow poison and had been mixed with the arrow poison. The heart stimulating effect was then verified in Europe and Fraser in England isolated k-strophantin in 1859 and recommended it and the seeds for medical use.

The seeds contain cardiac glycosides, phytochemically different, but similar for the three aforementioned species. The seeds of *S. gratus* contain 4–7% cardiac glycosides, which consist of 90–95% g-strophantin.

Ouabain is a similar glycoside isolated from the wood of species of *Acokanthera* as well as from *S. gratus*.

CO1A D Other cardiac glycosides

LILY OF THE VALLEY, *Convallariae herba* consists of the dried aerial parts of *Convallaria majalis*, (L.), family *Convallariaceae*, collected when the flowers are beginning to open, and contains 0.2–0.3% and in the flowers 0.4% cardiac glycosides. The main glycoside is convallatoxin (40–45% of total content) which is pharmacologically similar to strophanthin and also has a pronounced diuretic effect. Therapeutically a standardised extract is used in combination with other drugs for treatment of cardiac insufficiency.

OLEANDER, *Oleandri folium* The drug is the dried leaves of *Nerium oleander* L., family *Apocynaceae*, a decorative shrub in the Mediterranean region. The leaves contain about 1% cardiac glycoside with the main glycoside oleandrin with a rather weak cardiac but strong diuretic effect. Many cases of poisoning with this plant have been reported.

C01B Antiarrhythmics Class I to IV

C01B A Antiarrhythmics Class I The natural products quinidine and ajmaline belong to the group of blocking agents of the so-called 'rapid sodium channels' in the cell membrane.

QUINIDINE Quinidine is one of the quinoline alkaloids in *Cinchona* bark. (See P01, Antimalaria-drugs). Thus, quinidine is an antiarrythmic drug, whereas its stereoisomer, quinine, is an antimalarial. In antiarrythmically effective doses quinidine reduces the contraction capacity of the heart. The minute volume of the heart diminishes through its negative inotropic effect. Quinidine is used clinically for treatment of relapse into auricle fibrillation, and at extrasystolics and paroxysmal tachycardia and ventricular tachycardia.

AJMALINE The alkaloid ajmaline is isolated from the root of *Rauvolfia serpentina* and other *Rauvolfia species*, family *Apocynaceae* (see section C02, Antihypertensives). Ajmaline has a very similar mode of action to quinidine. It works through retardation of the speed of depolarisation and repolarisation as well as a retardation of the reactivation of the sodium-system. Its sedative effect is weak.

Ajmaline is used clinically in supraventricular and ventricular tachycardia and (mostly the ventricular) extrasystole.

Ajmaline has a poor oral resorption: that is why parenteral administration is preferred. There is no problem in simultaneous administration of ajmaline and digitalis.

C01C Cardiac stimulants (EXCL. cardiac glycosides)

DIHYDROERGOTAMINE This substance, which is used in the treatment of migraine as well as hypotension, is classified in C04 and N02C, Ergot alkaloids.

C01D Vasodilators used in cardiac diseases

This group comprises preparations used in ischaemic heart disease (when the coronary vessels are more or less narrow). On the other hand no drugs that increase the cerebral circulation are included. They are found in section C04, Peripheral vasodilators.

VISNAGA The drug, *Ammeos visnagae fructus*, consists of the dried, ripe fruits of *Ammi visnaga*, (L.), Lam., family *Apiaceae*, an annual plant from the Mediterranean region, which is cultivated in Egypt, but also in Argentina, Mexico and the USA. The fruit is very similar to that of the closely related species *Ammi majus* L., but can be differentiated with the aid of anatomical differences in the fruit wall.

The fruits of *Ammi visnaga* contain the pharmacologically active chromone (khellin) and pyrano coumarins (visnadin).

Khellin has a spasmolytic effect on smooth muscles in coronary vessels, in bronchi, and in the gastrointestinal, bile and urinary tracts, with the point of action in the muscle cell itself. Khellin has been used clinically in the treatment of angina pectoris and bronchial asthma, but its use has been limited by undesirable side reactions. Now it is used in combinations in low dose (5–10 mg/single dose). (See R03D.)

Visnadin has a coronary vasodilator effect, and the increased, circulation takes place without increased oxygen consumption.

Lactate, pyruvate and glucose in the venous blood are lowered after visnadin medication, while the content of free fatty acids is increased which shows that energy metabolism is favourably influenced by visnadin. The minute volume is increased without increase in the heart frequency and blood pressure is practically unchanged. Clinically visnadin is used in the treatment of myocarditis and angina pectoris.

C01 E Other cardiac preparations

HAWTHORN, *Crataegi folium cum flore, fructus.* A draft monograph on Hawthorn leaf with flower has been published by the European Pharmacopoeia which defines the drug as the dried whole or cut flower-bearing branches, up to 7 cm long, of *Crataegus monogyna* Jac., *Crataegus laevigata* (Poir.) D.C. (C. *oxyacantha* L.) or their hybrids or more rarely other *Crataegus* species. Hawthorn contains not less than 1.5% of flavonoids calculated as hyperoside.

The flavonoids constitute only one group of active compounds. Oligomeric procyanidins, based on the condensation of phenolic catechins to give molecules having from 2 to 8 monomeric units, constitute approximately 3% of the dry weight of the drug. A bewildering variety of *Crataegus* extracts based on differing plants, parts of the plant and solvents of different compositions and strengths have been tested for their cardioactivity. It is difficult to compare the results obtained with different dosages and different results. What is clear is that it is possible to use *Crataegus* in early forms of cardiac insufficiency and angina pectoris, in nervous heart complaints and to support cardiac and circulatory functions.

Pharmacodynamic tests *in vitro* and *in vivo* show that extracts, flavonoids and procyanidins have positive inotropic effects. An increase in coronary blood flow has also been demonstrated and it is the procyanidins that have the most significant activity. Antiarrythmic effects and slight hypotensive activity have been demonstrated largely due to the presence of procyanidins.

Possible mechanisms of action which have been suggested on the basis of *in vitro* studies include inhibition of Na+ - K+ - ATP-ase, and cAMP - phosphodiesterase activity, inhibition of thromboxane (TXA2) synthesis and an antioxidant effect. Clinical studies performed largely in patients in the early stages of cardiac insufficiency have shown significant improvements in both objective and subjective symptoms. Although *Crataegus* is usually used in patients who do not yet require *Digitalis* glycosides, there is a synergism between the actions of both and if a physician chose to prescribe both, the dose of *Digitalis* glycosides could be adjusted downwards by up to 50%.

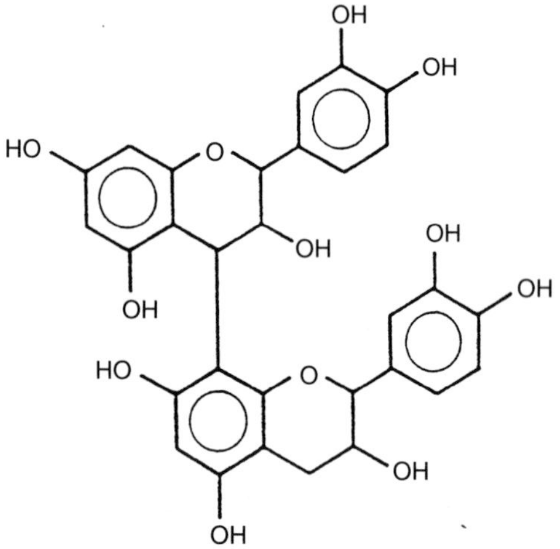

Procyanidin (dimeric flavonoid)

CO2 Antihypertensives

C02A Antiadrenergic agents, centrally acting

CO2A A Rauvolfia alkaloids

RAUVOLFIA-ALKALOIDS These have an interesting history. The Indian species *Rauvolfia serpentina* (L.) Benth. ex. Kurtz,

Fig. 12. *Rauvolfia serpentina* L. Benth. ex Kurz. Cultivation in India

family Apocynaceae, belongs to the very old traditional medicine (Fig. 12). Under the Sanskrit name Sarpagandha it was used in the case of snakebite, epilepsy, fever and mental illness. In the West it was the Augsburg physician and botanist Leonhart Rauvolf who described this plant in 1582 for the first time. Two Indian pharmacologists in 1933 described a long-lasting strong depressive effect on blood pressure for an alkaloid from the plant. This alkaloid was identified in 1952 by a Swiss research team as reserpine.

The root of *R. serpentina* was rapidly accepted by several pharmacopoeias. Later on, the pure alkaloids have been used in therapy and then other *Rauvolfia* species, mainly the African one, has been of current interest as the starting material for extraction of pure alkaloids. The *Rauvolfia* genus is strictly tropical. While the Asiatic species, *R. serpentina*, is a shrub, 0.5-1 m high, which grows in India, Bangladesh, Burma, Thailand and Java, the African species, *R. vomitoria* Afz., is a high shrub or tree of 10 m in height, growing in Africa from Guinea to Mozambique. *R. tetraphylla* (= *R. canescens*) growing in the Carribean region is used for extraction of pure alkaloids.

In the rootbark of *R. serpentina* more than 50 indole alkaloids occur. Only four have a hypotensive effect: reserpine, rescinnamine, deserpidine and serpentine.

H_3CO — N — H — H — H — OR — $H_3CO—C(=O)$ — OCH_3

	R
(–) – Reserpine	—OC— (ring with OCH_3, OCH_3, OCH_3)
(–) – Rescinnamine	—OC—CH=CH— (ring with OCH_3, OCH_3, OCH_3)
Deserpidine	H

The hypotensive Rauvolfia-alkaloids

These alkaloids can be considered as structural variations of yohimbine. Serpentine has a heterocyclic E-ring and is a strong base, whereas the others are weak bases.

According to their mode of action the *Rauvolfia-alkaloids* can be divided into two groups: the centrally acting reserpine, rescinnamine, and deserpidine and the peripherally acting yohimbine, raubasine and ajmaline. The blood pressure lowering effect is due to an inhibition of the sympathomimetic effects in the autonomic nervous system and has a long latency and duration. The effect is due to depletion of the serotonin and noradrenaline depots in the brain. High doses of reserpine can provoke psychotic tiredness and depression.

The hypotensive effect of reserpine is used in severe hypertension, nowadays in low doses in combination with other anti-hypertensives. Its sedative and relaxant effect has been used in the treatment of psychoses. Reserpine was the first useful remedy in the treatment of psychosis and opened up the road for modern psychopharmacotherapy, which has drastically changed the treatment of psychiatric patients.

CO3 Diuretics

All substances that increase the urinary flow are called diuretics. The diuretics which are most used therapeutically produce an inhibition of the tubular reabsorption of Na+ and Cl–, which gives an excretion of the electrolytes, which osmotically bind water and thus increase urine volume.

Among the plant derived diuretics, the xanthine derivatives have this mode of action, whereas the osmotic diuretics, mannitol and sorbitol, which produce a glomerular filtration, but not resorption in the tubules, incease the excretion of water.

XANTHINE-DERIVATIVES (see NO6) *Theobromine* has a stronger diuretic effect than caffeine and theophylline (RO3D).

MANNITOL AND SORBITOL (see BO5B C, BO5C) are mainly used in cases of cerebral and lung oedema.

Medicinal teas with diuretic effect are usually a mixture of different drugs. Here only four will be dealt with. In France there are officially 21 drugs accepted for this purpose, in Germany 28 drugs. These diuretics are 'water diuretics', also called 'aquaretics', and suitable for irrigation therapy used for dysuric symptoms, catarrh of the bladder, to prevent relapse in urinary infection and prophylaxis of stones (concrements).

JUNIPER BERRIES, *Juniperi fructus* Juniper berries are the dried ripe fruits of *Juniperus communis* (L.), family *Cypressaceae*, an evergreen shrub or small tree, growing in Europe. Generally speaking, the berries from the more southern contries contain the most essential oil (up to 2%), usually 1% with α-pinene as the main constituent and terpinen-4-ol as the diuretic constituent. The sweet taste is given by invert sugar. Beside the diuretic effect, it gives an increase in gastrointestinal secretion. By local application, it increases the peripheral circulation.

There have been suggestions made that Juniper berries could cause renal irritation, and that they might also be an abortifacient. In the former case it is probable that the Juniper oil suspected of causing irritation had been adulterated with turpentine oil thus increasing the pinene content. In the latter case it would appear that Juniper was confused with the related *Juniperus sabina* or savin which is a notorious abortifacient.

BIRCH LEAF, *Betulae folium* The drug consists of the dried leaves of *Betula pendula*, Roth, and *Betula pubescens*, Ehrh., family *Betulaceae*. The leaves have a weak aromatic smell and light bitter taste. They contain about 2% flavone-glycosides as hyperoside and quercitrin and also about 0.5% ascorbic acid. The diuretic and saluretic effect has been verified in animal experiments and may be due to potassium in the leaves.

ORTHOSIPHON LEAVES, *Orthosiphonis folium* The drug consists of the dried leaves including the tops of the branches of *Orthosiphon aristatus* (Blume) Miquel (syn. *spicatus* Thunberg) Backer and O. *stamineus* Bentham, family *Lamiaceae*. The plant is cultivated on Java ('Java-tea') and Cuba. The leaves have a weak aromatic smell and a light bitter and astringent taste. The leaves contain 0,2% lipophilic flavonoids and 3% potassium salt. Animal studies show positive effects but human studies have given contradictory results.

HORSETAIL, *Equiseti herba* The drug consists of the dried aerial part of the green sterile shoot of *Equisetum arvense*, (L.), family *Equisetaceae*. It contains organic silica, potassium, saponin, flavonoids and traces of alkaloids.

CO4 Peripheral vasodilators

CO4A Peripheral vasodilators

This group comprises single and combined preparations used in the treatment of cerebrovascular and peripheral circulatory disorders.

VINCAMINE Vincamine is an indole alkaloid, found in the leaves of the periwinkle, *Vinca minor* (L.), family *Apocynaceae*, indigenous to Central and South Europe, where it grows in shady places.

Clinical pharmacology shows that vincamine increases the cerebral circulation, and glucose uptake in ischaemic parts of the brain, and thus decreases cerebral oedema and the viscosity of the blood.

As compared to the effect of placebo, 60 mg vincamine administered orally daily gives a significant effect with improved intellectual performance.

GINKGO FOLIUM The drug consists of the dried leaf of *Ginkgo biloba* (L.), a tree belonging to the gymnosperms, in Japanese called icho. It is cultivated in France and S. Carolina (USA). A standardised extract of the leaves is used therapeutically. The main ingredients so far isolated from the leaves are:

1) Lipophilic diterpene lactones Ginkgolide A, B, C, D and E (ca 0.06%) and the sesquiterpene lactone Bilobalide (ca. 0.02%).
2) O-Acylglycosides derived from quercetin, isorhamnetin and kämpferol (ca. 10%).
3) Bioflavonoids (ginkgetin, isoginkgetin), catechin and procyanidin.

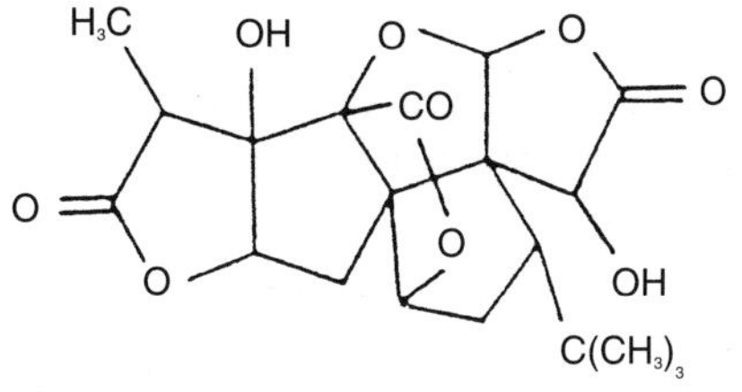

Ginkgolide A

Virtually all of the pharmacological and clinical results have been obtained with extracts standardised to contain 6% of terpenoids (the ginkgolides and bilobalide) and 24% of Ginkgo flavones. The individual components are known to have the following effects: the flavonoids (mainly quercetin, kampferol and isorhamnetin) have free-radical scavenging and antioxidant activity; the ginkgolides are powerful platelet activating factor (PAF) antagonists thus inhibiting platelet aggregation and reducing the viscosity of blood. These compounds in purified form are being tested for their anti-asthmatic effects; bilobalide reduces cerebral oedema in animals.

Over 100 clinical trials have been reported with *Ginkgo biloba* extract and in some cases sophisticated statistical meta-analyses of the trials have been performed. The most widely studied therapeutic use has been for a condition referred to as cerebral insufficiency which represents a collection of symptoms normally affecting the elderly such as difficulties in concentration, memory problems, confusion, dizziness etc. It is believed that reduced cerebral blood flow and oxygenation may be involved and that the Ginkgo extract improves blood flow, oxygenation and the harmful effects of free-radical formation due to hypoxia in neurones. Over 40 controlled trials have been subjected to independent review. The reviewers concluded that Ginkgo was of value in mild to moderate symptoms of cerebral insufficiency, noting that side-effects were low and that 4–6 weeks of continuous use was required to obtain any benefit.

The second most popular indication is intermittent claudication or disabling pain in the legs due to peripheral arterial disease. Two meta-analyses of the published trials have shown that Ginkgo gives a 45% increase in pain-free walking distance compared to placebo. Some studies have been performed in patients with dementia and reduced cognitive functions due to Alzheimer's Disease. A meta-analysis of four high-quality trials showed that 120–240 mg of Ginkgo extract daily for 3–6 months had a small but significant effect on cognitive function in 212 patients with Alzheimer's Disease, compared to placebo. The most recent such trial, published in the *Journal of the American Medical Association* in 1997, lasted for 1 year and involved 202 patients with either Alzheimer's or multiple infarct dementias. On one assessment scale twice the number of the Ginkgo patients showed improvement compared to placebo, while on another there was also a statistically significant improvement. It was concluded that Ginkgo extract was safe and capable of improving the cognitive and social functioning of patients with dementia. Ginkgo has also shown benefits in premenstrual syndrome, a 65% improvement in reaction tests in patients with subarachnoid haemorrhage and a 40% remission in cases of vertigo and sudden deafness.

HYDROGENATED ERGOT ALKALOIDS The parasitic fungus *Claviceps purpurea* Tulasne gives the drug ergot, *Secale Cornutum* (see G02A). Ergot contains three groups of alkaloids, one of them being the ergotoxine-group, consisting of three alkaloids, which all have a double bond in their lysergic acid part. When this double-bond is saturated, a semisynthetic alkaloid mixture is obtained. It is now called Co-dergocrine and consists of equal parts dihydroergocornine, dihydroergocristine and α- and β-dihydrocryptine.

Co-dergocrine has an α-receptor blocking and a centrally depressive effect. As distinct from the non-hydrogenated ergot alkaloids, the hydrogenated alkaloids in Co-dergocrine close the arteriovenous shunt and transform the blood so that a larger proportion goes to the capillary bed. Co-dergocrine is used as a vasodilator in the treatment of peripheral circulation disturbances, but mainly to relieve the symptoms of reduced mental capacity in the elderly through increased cerebral blood flow and lowered blood pressure.

CO5 Vasoprotectives

CO5A Antihaemorrhoidals for topical use

PERU BALSAM, *Balsamum peruvianum.* This is a pathological product, which is produced in El Salvador and other Central American countries from the bark of *Myroxylon balsamum* (L.), Harms var. *pereirae* (Royal) Harms, family *Fabaceae*,

The commercial product contains esters: benzyl cinnamate (cinnamein), benzyl benzoate, and cinnamyl cinnamate (styracin). The drug usually contains from 56 to 66% of these esters. Balsam of Peru is used as an antiseptic dressing for wounds (haemorrhoids).

TANNINS The tannin β-hamamelitannin in the leaves and bark of witch hazel, *Hamamelis virginiana* (L.), family *Hamamelidaceae*, has a mild antiseptic and astringent effect used for local treatment of haemorrhoids.

CO5B Antivaricose therapy

This group comprises all products for topical treatment of varices.

HORSE CHESTNUT, *Aesculi semen* is the seed of *Aesculus hippocastanum* (L.), family *Hippocastanaceae.* The seed contains a triterpene saponin mixture usually referred to as Aescin. Aescin gives prolonged antagonism against several types of local oedema, e.g. that caused by bradykinin, it increases capillary resistance and decreases capillary permeability. While Horse chestnut extracts have been used topically, clinical studies have shown it to be more effective in reducing leg oedema volume in patients with varicose veins when it has been administered orally.

BUTCHERS BROOM The roots of *Ruscus aculeatus* contain a mixture of steroid sapogenins called ruscogenins which are structurally similar to diosgenin. These have been formulated into topical preparations for varicose veins and haemorrhoids.

MUCOPOLYSACCHARIDE POLYSULPHATE This substance is absorbed percutaneously and exerts a local anti-inflammatory effect on superficial thrombophlebitis: in the treatment of infusion thrombophlebitis it has been shown - using the 125 - iodine fibrino gentest - that it produces a more rapid regression of the fibrin depot (the clot) in the area of inflammation. Also, the clinical symptoms were improved more rapidly.

CO5C Capillary stabilizing agents

In the group of *bioflavonoids*, there is rutin (=rutoside), which occurs in Fagopyrum (Buckwheat), Eucalyptus and *Solidago* species and in the flowers of *Sophora japonica*.

Rutin has an oedema protective and capillary stabilising effect. The effect on the capillaries takes place in the region of endothelial cells, with the result that the number of endothelial pores and their distribution is diminished. Most probably this effect depends on an inhibition of the activity of prostaglandin PGE_2.

In clinical trials rutoside clearly inhibits thrombocyte and erythrocyte aggregation and inhibits the symptoms of oedema.

D Dermatologicals

DO3 Preparations for treatment of wounds and ulcers

D03C Tannins

The tannins are a group of natural products, which in earlier days were used exclusively for tanning of skins to leather (today to a large extent substituted by inorganic salts: alum chrome-salts), but have a medical use owing to their astringent, protein-precipitating effect. On mucous membranes and wounds a coherent coagulation of membrane is formed.

As a consequence of this effect there is an astringent taste, dryness in the tissues, cessation of glandular secretion and small wounds. Redness, swelling and increased secretion in inflamed tissue are diminished. As a result of the removal of the coherent coagulant membrane in the mouth and throat, the enclosed pathogenic microbes are also removed and the formation of new epithelium is stimulated.

Preparations containing tannins are used externally in cases of damage by frost, haemorrhoids or small burns.

Internal use is in the treatment of catarrh of the stomach and as an antidiarrheal. Not all tannins are suitable for all uses. In the treatment of large burns, tannins are absorbed in such large quantities that they can cause serious damage to the liver. That is the reason why large burns are not treated with tannin solution any more.

There are two types of tannins: hydrolysable and condensed (*not* hydrolysable). The hydrolysable are more toxic than the condensed.

Hydrolysable Tannins These may be hydrolysed by acids or enzymes such as tannase. They are formed from several molecules of phenolic acids such as gallic and ellagic acids which are united by ester linkages to a central glucose residue.

The two main types of hydrolysable tannins may be derived respectively from gallic acid and ellagic acid (the depside of gallic acid). These are known as gallo tannins and ellagitannins.

GALLS, *Gallae* As a result of insect attacks on plants, pathological developments called galls are formed. Turkish or Syrian galls are formed on young twigs of *Quercus infectoria* Olivier, family *Fagaceae*, as a result of the depositon of the eggs of the gall-wasp. *Adleria gallae-tinctoriae*, which hatch into larvae which stimulate the surrounding tissue by means of enzyme-containing secretions produced by the insect to develop the gall, which is globular in shape. Chinese gall is a gall formation on *Rhus chinensis* Miller, family *Anacardiaceae*, produced by an aphid, *Melaphis sinensis.*

From these two galls, tannin of similar structure is extracted: one glucose molecule, esterified with a varying number of gallic acid-molecules.

Condensed tannins

OAKBARK, *Querci cortex* The drug consists of the bark of young branches and stems of *Quercus rubor (L.)* and *Quercus petraea* (Mattuschka) Liebl, family Fagaceae. The bark contains condensed tannins with polymers of catechin as the main constituents. Tannins of the same type are also found in rhattany root and tormentil root (see AO1A).

Cicatrisants

CALENDULA FLOWERS are the dried flowers of *Calendula officinalis L.*, family *Asteraceae.*

The flowers contain triterpenoid glycosides based on oleanolic acid, sesquiterpenes, carotenoids and flavonoids. Infusions and tinctures are used topically for inflammations of the skin and to promote wound healing (cicatrisation).

Antiinflammatories

CHAMOMILE FLOWERS *Matricaria recutita, L.*, family *Asteraceae*, contain an essential oil rich in sesquiterpenes such as bisabolol and chamazulene which have excellent antiinflammatory effects.

ARNICA FLOWERS are the flower heads of *Arnica montana* L, family *Asteraceae.* The flowers contain sesquiterpenes such as helenalin and dihydrohelenalin which have antiinflammatory and analgesic effects which explain the popularity of Arnica as a remedy for bruises and aches and sprains. It must be stressed that Arnica should not be used internally because of cardiotoxicity. Equally important is the fact that the sesquiterpenes, like most if not all of those from the *Asteraceae* (e.g. in Chamomile and Yarrow), are highly allergenic and patients must be advised to discontinue use if a skin reaction develops and to avoid Arnica and Chamomile if they are known to have a sensitivity to other *Asteraceae* such as Asters, Daisies and Chrysanthemums.

ALOE VERA is the name usually given to the gel which exudes from the cut leaves of *Aloe barbadensis*, Miller, family *Liliaceae* (Curacao or barbadoes Aloes) (cf. AO6). It is important not to confuse this mucilaginous gel obtained from the central part of the cut leaf with Aloes or Aloe resin (see A 06) obtained from the outer pericyclic cells of the leaf and subsequently concentrated by heat to give the laxative material. In contrast to the anthranoid content of Aloes, the gel contains mainly polysaccharides such as acemannan; pectins; the sugar, mannosose 6-phosphate; sterols; and an enzyme carboxypeptidase. Aloe Vera gel is widely used to promote wound healing and in the treatment of inflammatory conditions of the skin such as minor skin irritations and burns (including sunburn, bruises and abrasions). According to

WHO, most clinical investigations have found that Aloe vera accelerates wound healing through the effect of acemannan and mannose - 6-phosphate on fibroblasts. Many of the active ingredients are unstable and methods of stabilisation affect the activity. All fresh gel is known to be active and many stabilised gels have also been shown to be effective topically but others have been shown to actually damage cultured cells. The antiinflammatory effects are due to the inhibition of thromboxane B_2 and prostaglandin F_2 synthesis and involves phytosterols as well as enzyme activity. The effectiveness of orally administered gel is less well established compared to the evidence for topical administration. Positive results have been obtained in studies with diabetics, and acemannan has anti-HIV activity. Aloe Vera containing a minimum of 11.3 mg/ml of acemannan could be a useful adjunct in the treatment of HIV infection and AIDS. According to WHO internal use has not been shown to exert any consistent clinical effect, but antidiabetic, anticancer and antibacterial effects are receiving more attention, as are the immunomodulatory properation of the gel polysaccharides.

DO4 Antipruritics

MENTHOL Menthol is the active ingredient in *Menthae piperitae folium* (see A01) and a 1-2% alcoholic solution has an antipruritic effect.

THYMI AETHEROLEUM, is the active ingredient of *Thymi herba*, used as an expectorant (see RO5C) and which has also an antipruritic effect, verified in clinical studies.

DO5 Antipsoriatics

The most pleasant treatment is sun and salt baths and the effect is often of long duration. But most patients suffering from psoriasis must, at least during the greater part of the year, have dermatological treatment.

XANTHOTOXIN Xanthotoxin (8-methoxy - psoralen, 8-MOP) is found in the fruits of *Ammi majus*, (L.), family *Apiaceae*, which grows in the Nile delta, but is now also cultivated in India and Argentina. Xanthotoxin has two dermatological uses: psoriasis and vitiligo. The mode of action is in both cases an inhibition by ultraviolet light of DNA synthesis in the epidermis.

Xanthotoxin serves as a sensitising agent by absorbing the longwave UV-light and gives the absorbed energy to the cutaneous cells. In particular co-valent bonds are formed between xanthotoxin and the pyrimidine bases in the nucleic acids and cross-wise network in the region of the DNA double helix. The reduction in division of the nucleus gives the desired reduction of psoriasis in the increased growth of epidermis. Xanthotoxin is administered in doses of 20 mg and externally for local administration. The synthetic product 4,5′,8-trimethylpsoralen (TMP, trioxalen) is used externally.

For the treatment of *Vitiligo* (lack of pigmention in certain skin areas) two drugs are used: xanthotoxin and psoralen.

The last-mentioned substance is found in the fruits of *Psoralea coryfolia* (L.), family *Fabaceae*. Already in Papyrus Ebers (about 1550 BC) a treatment of vitiligo with the two plants was described when an extract of the fruits of *Ammi majus* was painted on the white spots of the skin, which was sunlit for some hours. A redness occurs, followed after 1-2 days by a strong brown colour. This is an example of photo-chemotherapy (drug+light).

TARS Tars, *Pyrolea*, are obtained by dry distillation of various trees. Thus, *Betulae pyroleum* is obtained by dry distillation of birch wood and *Fagi pyroleum* from beech wood. These tars are ingredients in ointments for local treatment of psoriasis.

DO8 Antiseptics and disinfectants

CHAMOMILE FLOWERS (*Matricaria recutita L.*), family *Asteraceae* Chamomile flowers contain chamazulene and bisabolol, whereas white willow, *Salicis albae cortex*, bark from *Salix alba*, L., family *Salicaceae*, contains salicylic acid and its derivatives. These substances have antiseptic and disinfectant effects.

THYME OIL distilled from the leaves and flowers of *Thymus vulgaris* L. or *Thymus zygis* L., family *Lamiaceae*, must contain not less than 0.5% of volatile phenols, calculated as thymol. This compound and the related carvacrol means that the oil is highly antibacterial and antifungal, showing activity against Gram-positive and Gram-negative bacteria, fungi and yeasts.

LEMONGRASS OIL, *Cymbopogon citratus* (D.C.), Stop, Family *Poaceae* and

CITRONELLA OILS *C. nordus* (L.), W. Watson and *C. winterianus*, Jowitt, are also powerful disinfectants with phenol coefficients of 17 and 10 respectively due to the presence of the monoterpenes gerianal and neral (α- and β-citral).

D10 Anti-acne preparations

HEART'S EASE, WILD PANSY, *Violae tricoloris herba* (L.), family *Violaceae*, has shown therapeutic effect in the treatment of acne.

TEA TREE OIL, is distilled from the leaves of *Melaleuca alternifolia* (COK), family *Myrtaceae.* It contains a number of antibacterial terpenes, including 1,8-cineole and terpinen 4-ol. A comparative clinical study of 5% Tea Tree oil and 5% benzoyl peroxide in acne patients for three months showed that Tea Tree oil was as effective as benzoyl peroxide in reducing the number of scars, but that it had a lower irritant effect on skin. Tea Tree oil has also been shown to have *in vitro* activity against 60 methicillin-resistant strains of *Staph. aureus* (MRSA) and was clinically effective in vaginitis and in chronic infections due to *Trichophyton* species.

G Genito urinary system and sex hormones

GO2 Other gynaecologicals

G02A Oxytocics

Plain preparations of oxytocin and derivatives are classified in H01B: Posterior pituitary lobe hormones. Oxytocin and derivatives.

G02A B Ergot alkaloids Ergot alkaloids have four different types of effects:

1) Oxcytocics used for stimulation of uterine contractions: Ergometrine, Methylergometrine
2) Peripheral vasodilators and other ergot alkaloids (Co-dergocrine) are classified in CO4A
3) Antimigraine preparations (Dihydroergotamine, Ergotamine, Methysergide) in N02C
4) Other psychostimulants (lysergic acid diethylamide) in Chapter VIII

ERGOMETRINE, METHYLERGOMETRINE. The naturally occurring alkaloid ergometrine is an acid amide of D-lysergic acid and 2-aminopropanol. The semisynthetic alkaloid is methylergometrine, in which the side-chain is 2-aminobutanol. The chemistry of ergot is dealt with in section NO2C.

Methylergometrine is used in childbirth to diminish bleeding but not to induce labour (when oxytocin is used and is given intravenously and the oxytocic effect appears after $\frac{1}{2}$–1 minute and has a duration of 4–6 hours).

Oral administration is used in the treatment of post partum haemorrhage, menorrhagia (excessive menstrual bleeding) and metorrhagia (non-menstrual bleeding).

G02A D Prostaglandins About 15 prostaglandins have been isolated from animal and humans (isolated from sperm), where they are biosynthetised from linoleic acid (from food) via arachidonic acid and a cyclic endoperoxide (from which prostacyclin and thromboxane are also formed). There are four series of prostaglandins based on the substitution pattern in the cyclopentane ring of the prostanoic acid: an unsaturated hydroxylated C20-fatty acid, which is part of the main skeleton of the molecule.

Prostaglandins have not yet been found in plants, but Prostaglandin A is found (about 1.5%) in the coral *Plexaura homomalla*.

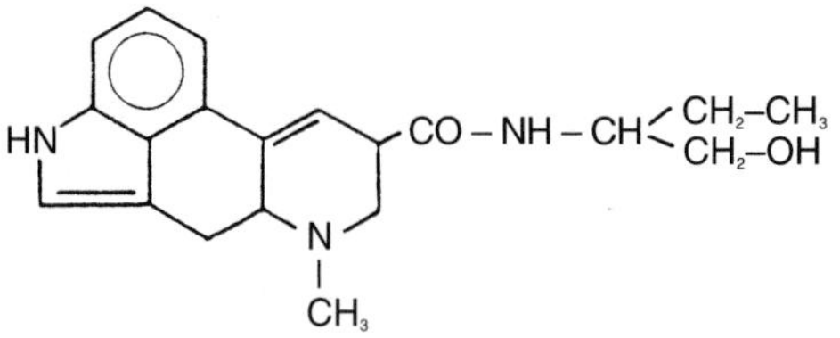

Methylergometrine

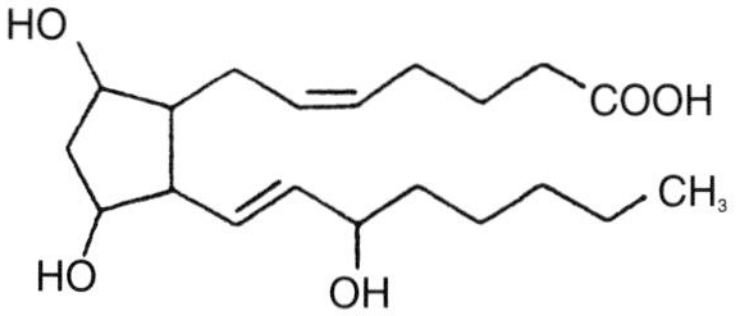

Dinoprost

(Prostaglandin F 2α)

DINOPROST (Prostaglandin F_2) The Prostaglandin F-series provokes contraction of bronchial, intestinal and pregnant uterine muscles. Therefore dinoprost is used clinically to terminate pregnancy in the first and second trimester.

G02A X Other oxytocics

RASPBERRY LEAVES, *Rubus idaeus* L., family *Rosaceae*, are stated to possess astringent properties and facilitate child birth. Uteroactivity has been documented in animal studies. It should not be taken during pregnancy unless under medical supervision.

G02C Other gynaecologicals

G02C B Prolactin inhibitors Bromocryptine is a semisynthetic ergot alkaloid, 2-bromoergo-cryptine and acts as an dopamine-agonist (i.e. has the same effect as dopamine) with a long duration and is an inhibitor of prolactin and growth hormone. For the indication inhibition and cessation of lactation one starts with a low dose: 2.5 mg taken with food for two weeks.

For treatment of acromegaly a starting dose of 2.5 mg is used, ending with 10–20 mg, depending on clinical and side-effects.

In the treatment of parkinsonism (see N04), characterised by nigrostrial lack of dopamine, bromocryptine (1.25–7.5 mg) is combined with L-dopa.

G02C C Antiinflammatory products for vaginal administration The flower of *Lamium album* L., (white deadnettle) Family *Lamiaceae* is used for local treatment as a bath or infusion for disturbances in the physiologic milieu of the vagina with increased discharge without pain and slight itching for one week. Another treatment consists of vagitoria containing lactic acid bacteria (*Lactobacillus acidophilus*) and lactose. The bacteria form lactic acid from lactose, which restores the normal acidic character of the vagina.

G02C D Menstrual irregularities Two main groups of disorders can be classified here:

1) *Dysmenorrhoea*, which is painful menstruation, *menorrhagia*, which is too excessive menstruation, and *metorrhagia*, which is a non-menstrual bleeding; the second group is:
2) Secondary amenorrhagia and oligoamenorrhagia.

Antidysmenorrhoica are drugs which are meant to alleviate pain of organic origin during menstruation. There are no specific drugs for this indication. Therefore, various plants with *spasmolytic, analgesic* effects are used, such as German chamomile (*Matricaria recutita* L.) Yarrow (*Achillea millefolium* L.) Lady's Mantle (*Alchemilla vulgaris*) and Raspberry leaves (*Rubus idaeus*).

For treatment of menorrhagia and metorrhagia the ergot alkaloids (ergometrine and ergotamine) are used (G02A B).

The emmenagogues (agents that promote the menstrual discharge) are used for the indication of secondary amenorrhagia and oligomenorrhagia.

The emmenagogue effect can be released by direct or indirect stimulation of the uterus.

Nowadays in practice only the indirectly acting, mainly skin irritating treatment such as hot bath and cataplasm are used. They are strongly hyperaemic and stimulate ovarian function and menstruation. Such indirectly acting plants are plants containing essential oils (*Thuja occidentalis, Juniperus sabina* (abortifacient action, when taken orally), *Myristica fragrans, Petrosilinum sativum, Tanacetum vulgaris*).

G02C E Premenstrual syndrome PMS – premenstrual syndrome – appears some days before the start of menstrual discharge with psychic and somatic symptoms: irritability, mood changes, headache, breast tenderness, feeling of bloating.

Low plasma levels of prostaglandins may be a cause of PMS, whereas women suffering from dysmenorrhea have a higher content of PGF2 in the endometrium than normal.

There is also a positive connection between the degree of pain and the amount of liberated prostaglandin in individuals, and the ratio PGF_2/PGE is increased; this ratio correlates best with the occurrence and severity of dysmenorrhoea.

In therapy three plants are used: Evening Primrose (*Oenothera biennis* L., family *Onagraceae*), Borage (also known as Starflower) (*Borago officinalis*, family *Boraginaceae*) and Black currant (*Ribes nigrum*, family *Grossulariaceae*). The seeds of these three species contain gamma-linolenic acid (GLA), which, as a precursor to prostaglandin, leads to increased endogenous production, which reduces symptoms of PMS. Gamma linolenic acid is also used in the treatment of atopic eczema.

G02C F Climacteric syndrome The climacteric syndrome becomes – as the name indicates – more or less pronounced when a woman reaches the climateric period (about 40–45 years of age) with cessation of ovarian function. As a result there is diminished oestrogen and progestogen production and simultaneously increased secretion of gonadotropic hormones.

Here the follicle-stimulating hormone (FSH) is definitely more increased than the luteinising hormone (LH), which leads to an inverse relation between the two gonadotrophic hormones. Furthermore, there is extraglandular synthesis of oestrogens from androgenous precursors in the ovaries and adrenal glands.

The syndrome includes vegetative-endocrine symptoms, metabolic dysfunctional symptoms and endocrine psychosymptoms; the most common symptoms are dysfunctional bleedings, hot flushes, palpitations, headache, and vaginal dryness. Since oestrogen receptors are found in osteoblasts, it has been accepted that osteoporosis in elderly women is connected with the drop in glandular oestrogen production.

In therapy substitution with oestrogens is used (hormone replacement therapy) and in phytotherapy two plants with hormone-like - but not substitution - effects are used.

CHASTE BERRY (*Vitex agnus castus L.*, family *Verbenaceae*). An extract of the fruits of this bush growing in the Mediterranean region gives an inhibition of FSH and a stimulation of LH and prolactin. However, investigations on rats have shown that the vitex-agnus-castus extract inhibits prolactin secretion, so that a dopaminergic effect is obtained. Clinical trials have indicated that *Agnus castus* may benefit women with PMS.

BLACK COHOSH (syn. Black snake root), *Cimicifuga racemosa, Nutt.*, family *Ranunculaceae*

Contains a mixture of saponin-like triterpene-glycosides, including actein, 12-acetylactein and cimigoside. The endocrine effect of the extract is a selective reduction of the serum concentration of LH, whereas the FSH and prolactin concentrations are unchanged.

G03 Sex hormones and modulators of the genital system

G03A hormonal contraceptives for systemic use

Oral contraceptives for women contain as active ingredients progestogens and oestrogens. These substances are semisynthetic derivatives, which are based on naturally occurring steroids.

Contraceptives for men are few. Gossypol is extracted from cotton seed oil and tested in China. The difficulty is to find a fully reversible substance.

G04 Urologicals

G04A Urinary antiseptics and antiinfectives

Only non-complicated infections can be treated with phytotherapy, often in combination with diuretics.

The plants of the family *Ericaceae* are used therapeutically. The active principle is hydroquinone, which is the hydrolytic product in alkaline urine of the glycosides arbutin and methylarbutin (both inactive), which occur in the two *Ericaceae* drugs:

UVA URSI, Bearberry Leaf, *Uvae ursi folium, Arctostaphylos uva ursi* (L.) Spreng, family *Ericaceae.* The dried leaves constitute the drug. Besides arbutin (not less than 8%) and methylarbutin, the leaves also contain tannins. 1.5 - 4.0g dried leaves given by infusion three times daily gives an antiseptic effect (by hydrolysis to hydroquinone) in the urinary tract, when the urine is kept alkaline. The maximal antibacterial effect is reached about 3-4 hours after administration.

In order to avoid side-effects, the daily administration should not exceed 2 weeks and should involve the use of cold water to prepare an infusion with a low content of tannin.

OH Enzymatic hydrolysis Intestine → OH / OH Resorption Conjugation → OR / OH Hydrolysis pH > 8 → OH / OH

R = $-SO_3H$, –GluA

Arbutin (in effective precursor) Hydroquinone (antimicrobiologically effective)

COWBERRY (Cranberry), *Vitis idaeae folium*, from *Vaccinium vitis idaea* L. The leaves have the advantage of having a lower content of tannins. The arbutin content is about 5%.

CRANBERRY Two species of *Vaccinium* are known by this name. The North American species is *V. macrocarpum* Aiton, while that from Europe is *V. oxycoccos* L. For many years the drinking of cranberry juice was recommended as a way of reducing bacterial infections in the urinary tract because of the acidic urine produced. More recent research shows that cranberry juice is effective in treating but more importantly preventing urinary tract infections, because it inhibits the adherence of *E. coli* to the epithelial cells lining the urinary tract. This means that the conditions for the multiplication of bacteria are reduced. One of the components of the juice with antiadherence effects, is believed to be fructose but the main components appear to be proanthocyanidins, which are structurally similar to the condensed tannins. Clinical studies show that 300 ml of juice per day can markedly reduce bacteriuria and pyuria in elderly women.

SANDALWOOD, *Santali albi lignum*, is the heartwood of *Santalum album* L., family *Santalaceae*, a tree in South-East Asia. The essential oil, *Santali aetheroleum*, is produced by distillation of the wood under high pressure. This oil is used as a bactericide in urinary infections. Its mode of action is most probably a conversion in the body of α, β - santalol to the corresponding acids. Furthermore, it has a spasmolytic action. It is also used as an antiseptic and fragrant ingredient in cosmetics.

GO4B Other urologicals incl. antispasmodics

GO4B C Urinary Concrement Solvents

THE MADDER, *Rubia tinctorum*, L., family *Rubiaceae*, gives the drug *Rubiae tinctori radix* containing ruberythrinic acid (alizarin glycoside), which has anti-inflammatory, diuretic and spasmolytic effect on the urinary tract. By chelating it inhibits, in acidic urine (pH 6-6.5), the formation of calcium phosphate and calcium-oxalate stones. It has been used in cases of renal calculi, also as a prophylactic. It is administered together with magnesium salts. The dosage is optimal when the urine is pink coloured. It has been placed on an EU negative list because of links between anthraquinones and tumours.

G04B E Urologicals used in benign prostatic hypertrophy

Benign Prostatic Hypertrophy (BPH) is characterised by micturition problems, i.e. dysuria, polliakisuria, delayed start of micturition and incomplete emptying of the bladder.

BPH can be diagnosed in about 90% of men over 65 years of age but only in 30-40% are there clinically obvious troubles at micturition.

BPH is divided into 4 stages (according to Vahlensieck, 1985):

Stage I No micturition disturbances. More or less pronounced BPH.
Stage II Varying micturition disturbances. Urine flow 10-15 ml/sec. and no or low residual urine (up to 50 ml).
Stage III Permanent micturition disturbances, more or less pronounced BPH. Residual urine more than 50 ml.
Stage IV Permanent micturition disturbances, more or less pronounced BPH. Residual urine more than 100 ml. Dilation of the bladder.

The exact cause of BPH is not yet known. Probably there is a multifactorial background. Consequently there are several hypotheses. One of them, the dihydrotestosterone (DHT) hypothesis, claims that DHT is formed in the prostate from testosterone by the enzyme 5 α-reductase and is deposited in prostate tissue, causing the proliferation of the prostate. The concentration of DHT was found to be 5 times higher in BPH than normal. Based on these facts, drugs with a selective inhibition of 5 α-reductase are sought.

Phytotherapy has given objective therapeutic results in stages I-III of BPH. Other therapeutic strategies could include the use of eicosanoid inhibitors, and a reduction in sex hormone binding globulin (SHBG) binding capacity.

SAW PALMETTO is the fruit of *Sabal serrulata* (Mich.) Benth et Hook. (syn. *Serenoa repens* (Bart.) Small), family *Arecaceae*. The fruits contain fatty acids and steroids, particularly β-sitosterol, and liquid CO_2 hexane extracts of the berries are commercially available. Pharmacological studies have shown inhibition of 5-α-reductase and a reduction in hormone receptors in prostate tissue. Over 18 clinical trials involving nearly 3,000 men have been published. A meta-analysis of the best studies showed that patients given saw palmetto extract had a lower frequency of nocturia and an improvement in peak urinary flow compared to those on placebo. Side-effects were much lower than with synthetic medication such as finasteride.

NETTLE ROOT, (*Urtica radix*) is obtained from *Urtica dioica* L and *Urtica urens* L., family *Urticaceae*, and contains a lectin called *Urtica dioica* agglutinin, polysaccharides, sterol glucosides (especially β-sitosteryl-glucosides), phenyl propanes and lignans. Like sabal, nettle root and its extracts are recommended for the treatment of the symptoms of the micturition problems associated with BPH. Pharmacological studies show inhibitory effects on SHBG binding, on aromatase (an enzyme which converts testosterone to oestrogens), on Na^+/K^+ -ATP-ase activity in BPH cells and on growth of human BPH-tissue cells.

Studies in humans confirm the *in vitro* tests while clinical trials show clear benefits in terms of increased micturition volume, decreased frequency, decreased nocturia and a decrease in prostate volume.

PRUNUS AFRICANA (Hook. f.) Kalkm. (syn: *Pygeum africanum* Hook f.), family *Amygdaleceae*, is the bark sourced from various African countries traditionally used for the treatment of bladder problems. Constituents of the bark include phytosterols (β-sitosterol glucosides), triterpenes, ferulic acid esters and long-chain alkohols, particularly docosanol.

Pharmacological studies have shown that bark extracts inhibit prostaglandin synthesis in prostatic tissue, reduce cholesterol and reduce oedema due to inflammation in addition to effects on growth factors involved in the pathogenesis of BPH. Numerous clinical trials have demonstrated benefits in patients with BPH. Prostate size was reduced, as were residual urine volume and the number of episodes of nocturia. A review of clinical data from over 2,000 patients published over a 25-year period showed that *P. africana* bark was an effective drug for the treatment of the symptoms of mild to moderate BPH. Reported side-effects were low.

POLLEN The active ingredients are believed to be short-chain peptides, phytosterols and glycopeptides but their exact identity has not been confirmed. By incubation of prostatic-carcinoma-cell cultures with pollen, a diminished cell proliferation has been observed. In a multicentre, double-blind study all relevant parameters showed significant improvements after administration of pollen extract.

H Systemic hormonal preparations, excl. sex hormones

HO2 Corticosteroids for systemic use

The industrial production of glucocorticoids, mineralocorticoids, androgens, oestrogens and progestogens (see G03A) is connected with each other both scientifically and historically; therefore it is dealt with here briefly.

In 1929 oestrone was the first steroid hormone isolated from the urine of pregnant women. In 1934 the isolation of progesterone from ovaries of pigs was reported by no lower than four research groups independently of each other. In

1936 Reichstein in Switzerland and Kendall in the USA succeeded in isolating the glucocorticosteroid cortisone in small quantities: 200 mg cortisone from the adrenals of 20,000 cattle. Ten years later this hormone was available in greater quantities by partial synthesis via desoxycholic acid from bile as the starting material (however, with low yields in all synthetic steps). Huge clinical interest arose in 1949, when Hench in the USA reported dramatic effects of cortisone in the treatment of rheumatic arthritis. This led of course to a big demand for cortisone, which could not be fulfilled by the methods of synthesis available at that time.

The primary problem in the synthesis of cortisone was to find a suitable starting material containing a hydroxy or ketogroup in the 11-position of the steroid skeleton, because at this time there was no synthetic method available to introduce these groups in this position. There was a feverish search for 11-hydroxylated steroids occuring in nature. Prof. Tadeus Reichstein at the Pharmaceutical Institute in Basle had started a systematic research on cardiac glucosides (described in section CO1A), and from the seeds of the liane *Strophanthus sarmentosus* D.C., family *Apocynaceae*, he isolated the aglycone sarmentogenin with the desired hydroxy-group in the 11-position.

A veritable battle over this substance in Western Africa was started by different institutions, vividly described in the book *Green Medicines* by Margaret B. Kreig. However, sarmentogenin did not become the substance that one had hoped, because this species is highly polymorphous, giving big variations in the glycoside composition. The solution of the problem came from Mexico, where the American chemist Russell E. Marker was active. He had discovered that steroid sapogenins could be degraded to progesterone. In an inventory of the flora of Mexico, he discovered that there were many plants that contained suitable saponins in large quantities.

Based on Marker's two discoveries, the well-known company Syntex was started. After some time, however, Marker left Syntex and took all 'know-how' about the important synthesis with him. Fortunately, the Hungarian chemist George Rosenkrantz succeeded in solving the problem on the crucial synthesis of progesterone, and thus Syntex became one of the world's biggest producers of steroid hormones, because methods to convert progesterone into other steroid hormones, including cortisone, were found. To introduce a hydroxy-group in the 11-position microbiologically was a very important achievement (the figure shows the formation of 11-progesterone with *Rhizopus*

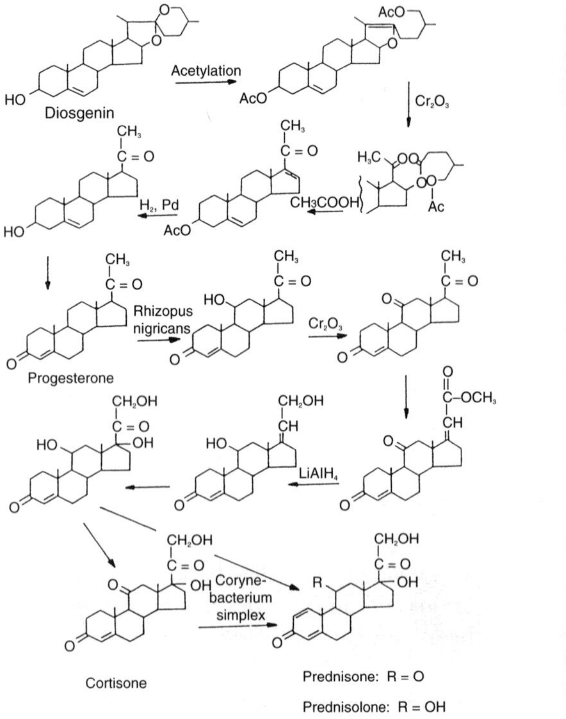

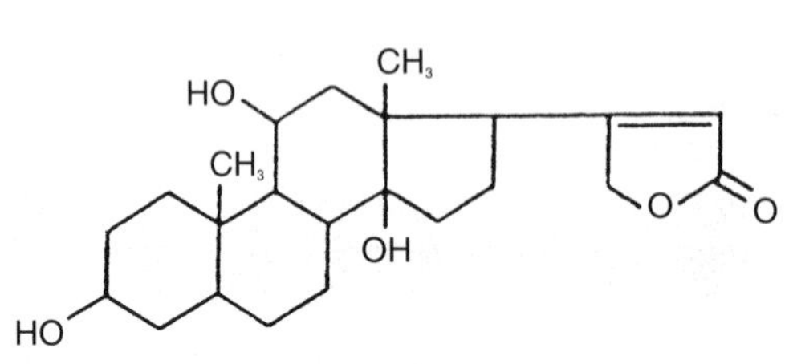

Sarmentogenin

nigricans). Among the steroid sapogenins, diosgenin has been the most useful starting material for the synthesis.

Precursors for steroid hormone synthesis

DIOSGENIN Diosgenin is a saponin aglycone, which is obtained from *Dioscorea* species, family *Dioscoreaceae.* The collection of roots from wild plants is carried out in Mexico from 'Barbasco' roots from *D. composita* Hemsley, by the rural population during the dry season (Fig. 13). The newly harvested roots are transported to a 'beneficio'. First, the roots are washed with water and converted to a thick paste, which is left to ferment. Then it is dried and sent to an extraction factory, where an acidic hydrolysis is performed, and the crude diosgenin is extracted with hexane.

Harvesting of cultivated plants of *Dioscorea floribunda* Mart.et Gal. in Bangalore is carried out analogously with the important difference that the whole root system is harvested after 2 years compared to the 3–5 years necessary for wild plants in Mexico.

Fig. 13. *Dioscorea composita* Hemsley

New sources for diosgenin are constantly being developed in the tropics. A promising species is *Costus speciosus* (J.G. Koenig), Smith, family *Costaceae*, which now is cultivated in India and Thailand, while *Trigonella foenum – graecum* has received attention in countries with temperate climates.

HECOGENIN Hecogenin is a steroidal sapogenin, which can be obtained from the waste juice from the preparation of sisal from the leaves of *Agave sisalana* (Engelm.) J.R. Drumm and Prain, family *Agavaceae* (Fig. 14). The plant is a succulent with long, fleshy, water-storing leaves, which have long tenacious fibres, from which sacks, ropes and other things are manufactured. Sisal is cultivated in Tanzania, Kenya and on the Yucatan peninsula.

The press juice is fermented in open barrels, and air is blown through the juice. 0.5% phenol is added to precipitate the crude hecogenin, which is filtered and dried and sent to the steroid factory, where the existing keto-group in the 12-position can be transferred via bromination to the 11-position and further to the various cortisone derivatives.

Fig. 14. *Agave sisalana* (Engelm), J.R. Drumm and Prain. Cultivation in Tanzania

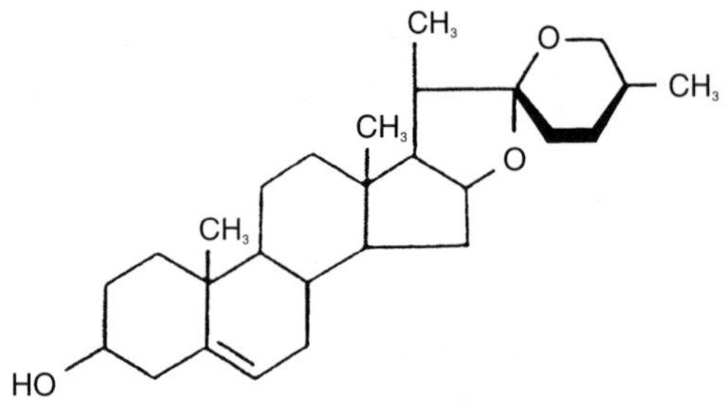

Diosgenin

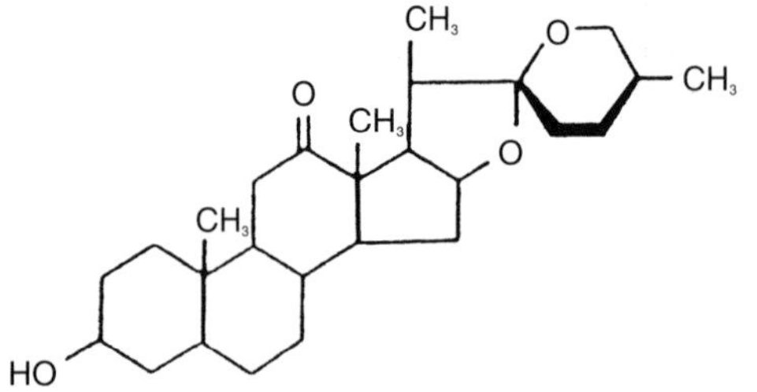

Hecogenin

SOLASODINE Solasodine is the N-analogue (in the F-ring) of diosgenin. It is obtained mainly from *Solanum laciniatum* Aiton, family *Solanaceae*, in Chimkent in Kazakstan by extraction of the fresh aerial part of the plant. In India a thornless mutant, *Solanum khasianum* C.B. Clark, is used for extraction of solasodine. The same scheme of synthesis as used for the diosgenin can be used for production of steroid hormones.

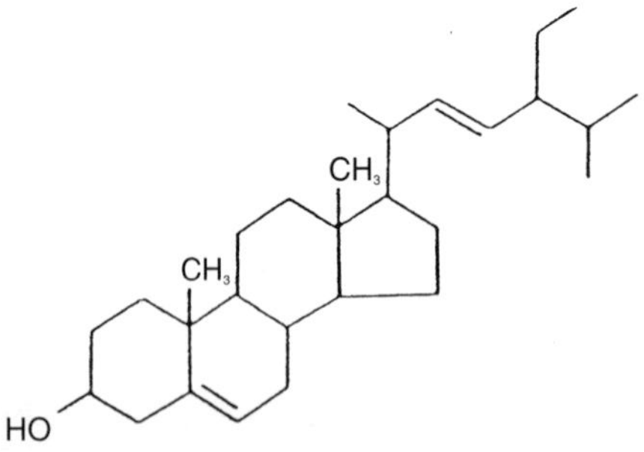

STIGMASTEROL Stigmasterol is a phytosterol which can be obtained industrially from the unsaponifiable part of soy oil, which contains 12–25% stigmasterol. At present, stigmasterol is the most used starting material for the production of steroid hormones. However, the chemical method for the transformation of stigmasterol to the intermediate progesterone is different from the corresponding transformaton of diosgenin.

SITOSTEROL Sitosterol is another phytosterol and is an analogue of stigmasterol with a saturated side chain. It is obtained from soy oil and corn oil. The side chain cannot be degraded chemically, but microbiologically it is possible to produce intermediates of value to the steroid industry. Sitosterol can therefore be used as a starting material for the synthesis of steroid hormones.

Also, cholesterol from the grease of wool (*Adeps lanae*) and bile acids from slaughtered food animals can be used as starting materials for the synthesis of steroid hormones.

L Antineoplastic and immunomodulating agents

L01 Cytostatics

L01C Plant alkaloids and other natural products

The Catharanthus alkaloids, taxol, colchicine and podophyllum derivatives have specific points of attack in the cell cycle and are phase-specific.

L01C A Catharanthus alkaloids and analogues

Catharanthus roseus (L), G. Don, (syn. *Vinca rosea* (L.), *Lochnera rosea* (L.), Reichb.), family *Apocynaceae*, is a perennial herb, 50–80 cm high. The flowers are white, pink or red. The plant originates from Madagascar and is now pantropic in its distribution. It is cultivated for two purposes: as an ornamental plant and for extraction of alkaloids (Madagascar, India, Sri Lanka). (Fig. 15)

Fig. 15. *Catharanthus roseus* (L.), G. Don (= *Vinca rosea* L. = *Lochnera rosea* L. Reichb.) Cultivation in the Tuticorin district, Southern India.

No other medicinal plant has been dealt with so extensively in the literature as this plant. More than 9,000 publications deal with this species from botanical, phytochemical, pharmacological and clinical points of view.

Up to now about 80 alkaloids have been isolated. Some of them exert a hypoglycaemic effect, verifying their use in traditional medicine as an antidiabetic drug. During the investigation of the hypoglycaemic effect of some alkaloid fractions it was found that they had decreased the number of leucocytes in the circulating blood and there was also an inhibition of erythropoesis in the bone marrow. This important observation led to the discovery of the tumour inhibitory effect. Four (two natural and two semi-synthetic) alkaloids are used in the treatment of definite forms of tumours. These alkaloids occur in the leaves in very low concentration. To obtain 1 g of pure alkaloid, about 500 kg of dried leaves must be extracted.

The monomers catharanthine and vindoline, which form the dimer, occur in much higher concentration, and it has been possible to synthesise the dimeric vinblastine (sterically correct) from the two monomers.

The Catharanthus-alkaloids inhibit mitosis, binding to the nucleus protein tubulin in the metaphase. There are differences among the alkaloids as regards pharmacokinetics, activity spectrum and toxicity.

VINBLASTINE is administered i.v. once weekly in doses of 6 mg/m^2 (up to 18 mg/m^2 body area) in the treatment of Hodgkin's disease and choriocarcinoma.

VINCRISTINE is given i.v. once weekly in a dose of 1-2 mg/m^2 in combination with other cytostatics in the treatment of leukaemia, malignant lymphoma and lung cancer.

VINDESINE is given i.v. in a dose of 3 mg/m^2 once weekly in the treatment of lung cancer (not the type with small cells), malignant melanoma and acute lymphatic leukaemia in children.

VINRELBINE Which is semisynthetic, is administered i.v. in a dose of 25–30 mg/m^2 once weekly. It is also used in combination therapy.

Vinblastine : R = $-CH_3$

Vincristine : R = $-CHO$

Fig. 16. *Podophylum hexandrum* Royle (= P. emodi Wallich ex Hook f. et Thomsen)

L01C B Podophyllotoxin-derivatives The drug *Podophyllin* is obtained by precipitation of an alcoholic extract of the rhizome of Mayapple, *Podophyllum peltatum L.*, family *Podophyllaceae* and *Podophyllum hexandrum*, Rogle, with acidic water. (Fig. 16) After drying, a yellow amorphous powder with an astringent and bitter taste is formed. *Podophyllin*, which occurs in the rhizome, (3-6% w/w) contains about 20% podophyllotoxin, 10% α-peltatin and 5% β-peltatin. These compounds are lignans derived from two phenylpropanoid units. The strong laxative effect is mainly given by the peltatins, which in low dose increase motility and secretion in the colon. Both peltatin and podophyllotoxin have a tumour inhibitory effect, which is exploited in the use of podophyllin dissolved in liquid paraffin, which has long been a topical treatment of *Condylomata acuminata*, which is a kind of contagious wart on the outer genitalia. Warts on the hands may also be treated in the same way.

This inhibitory effect on mitosis has been studied further and has led to two semisynthetic derivatives Etoposide and Teniposide, where a methylgroup in Etoposide is changed to a thiophene-group in Teniposide. The semisynthetic derivatives have a premitotic effect, and stop the inclusion of thymidine and uridine in the DNA and RNA in cell and protein synthesis.

Etoposide is administered orally and parenterally in lung cancer (small cells), acute non-lymphoblastic leukaemia and also in Hodgkin's disease, whereas Teniposide is given by intravenous infusion in Hodgkin's disease, and certain forms of cerebral tumours and cancer of the urinary bladder.

Etoposide and Teniposide have a cytostatic effect in the S-phase and G2-phase of the life cycle of the cell and differ from other podophyllin derivatives by not causing accumulation in the metaphase but prevent the cell from mitosis or destroy the cells that undergo mitosis.

Podophyllotoxin

Tenoposide

L01C C Colchicum and demecolchicine Colchicine is an alkaloid, previously extracted from the seeds of the Autumn Crocus or Meadow-saffron, *Colchicum autumnale* L., but nowadays from the seeds of *Gloriosa superba* L., both belonging to the family *Colchicaceae*. While *Colchicum* *autumnale* is a rare plant in central Europe, *Gloriosa superba* is a rather common climber in South India and Sri Lanka.

Colchicine is built from an aromatic ring with three methoxy groups, a 7-membered ring, to which is bound an acetylated aminogroup and a tropolone-ring with a methoxy group. Since the nitrogen is amide-bound, colchicine lacks the properties of a base. Demecolchine is N-desacetyl-N-methylcolchicine.

Colchicine is the oldest known mitosis-inhibitor isolated from plants. The cytotoxic effect depends on its binding to the nucleus protein tubulin in the metaphase. It has been used therapeutically in the treatment of acute gout. (See also section M04A.) Colchicine inhibits cell division in metaphase and has thus a cytostatic effect.

Colchicine has, however, a very small therapeutic index; a better one occurs in demecolchicine, where the toxicity is reduced 30–40 times. In a dose of 5–8 mg it is used in chronic myeloid leukaemia and in malignant lymphoma (lymphosarcoma).

L01C D Taxanes

TAXOL® (Paclitaxel) was first isolated from the bark of the Pacific Yew, *Taxus brevifolia* Nutt., family *Taxaceae*. Because of its value, supplies were difficult to obtain. Attempts to address the supply issue have involved semisynthesis from baccatin III or 10-deacetylbaccatin III which are taxanes found in the needles of the common European Yew *Taxus baccata* L. Other approaches have involved the use of fungal and tissue culture.

Taxol® acts by preventing the transition from metaphase to anaphase during mitosis through the polymerisation of tubulin, this including stable non-functional microtubules. This is a different mechanism of action from those of other plant antitumour substances. Taxol® is used in case of ovarian cancer which has spread to other organs (metastasised) and which is resistant to other chemotherapeutic agents. It is also useful in breast cancer under similar circumstances. The recommended dose is 175 mg/m^2 given as an intravenous infusion over a period of three hours. The formulation is difficult because taxol is hydrophobic and must be solubilised.

Docetaxel (Taxotere®) is a semisynthetic more water-soluble derivative produced from 10-deacetylbaccatin III. It is slightly more active than paclitaxel as a promoter of tubulin polymerisation and more active as an inhibitor of cell replication. Clinically the pattern of use of the two taxoids is similar.

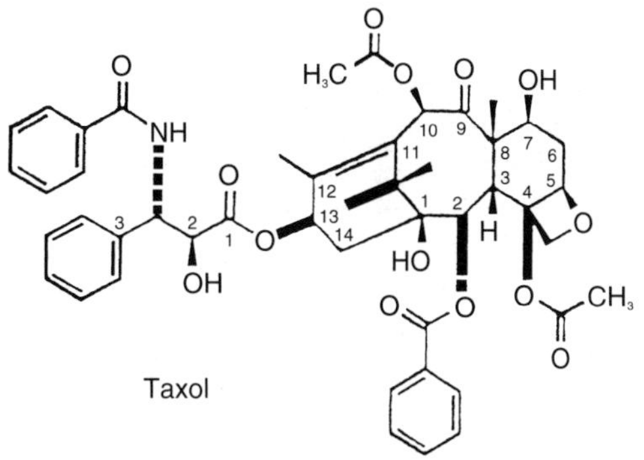

Taxol

Campothecin is an alkaloid from *Campotheca acuminata*, Decne, family *Nyssaceae*, a tree native to China and Tibet. It showed activity in antitumour screens but also unacceptable toxicity such as severe and unpredictable myelosuppression. Chinese researchers showed that it was possible to reduce toxicity by reducing particle size and by combining the drug with ammonium glycyrrhizinate (from liquorice) and used it in cases of liver, head and neck-cancers. Western interest was revived when it was discovered that campothecin and its derivatives had a novel mechanism of action involving selective inhibition of the enzyme topoisomerase I which is involved in the cleavage and reassembly of DNA. This led to the development of two semisynthetic derivatives *Irinotecan* and *Topotecan* which have been introduced clinically. *Irinotecan* is used in advanced colorectal cancer with clinical studies showing a prolongation of survival and an improvement in the quality of life. Topotecan has been used to treat metastatic solid tumours.

L01C E Mistletoe preparations

MISTLETOE, *Viscum album* L., family *Viscaceae*, is a semi-parasite, growing on leafy trees. There are other subspecies that grow on coniferous trees. The mistletoe takes water and salts from the host plant, but it has its own photosynthesis and its secondary metabolites are totally different from those of the host plant. Five main groups are said to be characteristic of mistletoe:

- polypeptides (viscotoxins A2, A3, and B)
- basic proteins
- glycopeptides (Lectins I, II, III)
- polysaccharides (acidic arabinogalactans)
- lowmolecular phenolic substances (flavonoids, organic acids, phenylpropanes and lignans)

It has been suggested that the lectins and viscotoxins are responsible for the antitumour activity. The viscotoxins (m.w. about 5,000) inhibit cell growth, but are 100 times less active than the lectins.

The mistletoe lectins (m.w. 60–120,000) have, in very low doses, an immunostimulating effect. The F-lymphocyte stimulant effect of several plant lectins is known.

Both *in vitro* and *in vivo* results have shown unanimously that the antitumour effect of mistletoe extract in low doses, corresponding to 0.25 to 1 mg lectin/kg, only depends on an immunoinducing mechanism of action.

The preparation Helixor® contains aqueous extracts of fresh mistletoe from 3 subspecies, namely *abietus, malus* and *pinus*, which have been sterilised by low temperature filtration. This product is administered subcutaneously. In low doses Helixor® has an immunostimulating effect; in higher doses it exerts a cystostatic effect, which is associated with the labile, basic proteins. Other investigations, have suggested that a complex polysaccharide is the active material.

L03 Immunomodulating agents

L03A Immunostimulating agents

There are many reasons for immunodeficiency states; change of lifestyle; alcohol abuse; severe trauma, severe metabolic disease (e.g. diabetes), acute infections, among others.

Immunostimulation is defined as the non-specific enhancement of cellular and/or humoral immune reactions. Immunostimulants do not produce antibodies but do lead to changes in the immune system such as increased formation of granulocytes and macrophages, increased phagocytic activity as well as the following:

- increased proliferation of T-lymphocytes
- increased activity of Natural Killer-cells
- increased liberation of mediators from monocytes and lymphocytes (monokines and lymphokines, such as interleukins, interferons, colony stimulating factor (CSF), tumour necrosis factor (TNF) prostaglandins)
- increased complement activity
- fever and increased liberation of inflammatory substances (prostaglandins).

After oral administration the immunostimulating substance can be in interaction with the pharyngeal-orientated immuno system in the tonsils.

ECHINACEA Three different species are used, i.e. *Echinacea angustifolia* (DC) Heller, *Echinacea pallida* (Nutt), Britt., and *Echinacea purpurea* Moensch, family *Asteraceae*. These species are cultivated in Europe. The herb, root or fresh juice pressed from the aerial parts may be used.

The chemistry of Echinacea is well documented and varies with the plant and plant part used. *Echinacea purpurea* herb contains three major groups of compounds, namely:

1) alkamides which are isobutylamides of straight chain fatty acids with 11 to 16 carbon atoms;
2) caffeic acid derivatives of which cichoric acid (2,3-0-dicaffeoyl - tartaric acid) is the most important;
3) polysaccharides including a glucurono-arabinoxylan (M.W. 45,000 D) and an acidic arabinorhamnogalactan (M.W. 450,000 D.).

The root of the purple coneflower *E. purpurea* contains similar compounds as well as polyacetylenes and glycoproteins, while the root of *E. pallida* (pale coneflower) contains caffeoyl derivatives such as echinacoside and ketopolyines together with small amounts of polysaccharides and glycoproteins.

Both *E. angustifolia* (which is of lesser commercial importance) and *E. purpurea* have been reported to contain saturated pyrrolizidine alkaloids. These, however, do not present any risk of liver toxicity because of the low concentration (0.006%) and because they are saturated molecules. The sesquiterpene lactone esters reported in *E. purpurea* have subsequently been attributed to *Parthenium integrifolium* which is a known adulterant of *E. purpurea.*

PHARMACOLOGICAL EFFECTS The type of extract or preparation can be important since alcoholic extracts are unlikely to contain polysaccharides, whereas aqueous extracts may not contain the lipophilic alkamides. Tests have been performed with the pressed juice, alcoholic extracts and isolated pure compounds. Effects such as increased phagocytosis as measured by the carbon clearance test, macrophage activation, and increased synthesis of TNF, interleukin 1 and interferon β have been reported. The alkamides and cichoric acid show strong effects on phagocytosis. Injection of polysaccharides into mice provides protection against lethal infections due to *Candida* and *Listeria.* In addition to its immune stimulant activity cichoric acid also inhibits hyaluronidase and thereby helps prevent the spread of infection. Recently it has been reported to have anti-HIV activity. Another phenolic component, echinacoside, has antibacterial and antiviral effects while the alkamides have inhibitory effects on lipoxygenase and cyclooxygenase. Polysaccharides from tissue cultures of *Echinacea* have also shown significant effects on macrophages, granulocytes and on cytokine production.

Clinical trials have been conducted on many *Echinacea* products administered both parenterally and orally. Trials have shown significantly enhanced phagocytosis peaking after 4–5 days. In another trial *Echinacea* juice was combined with econazole in the treatment of vaginal candidiasis. In the case of the combined products the recurrence rate of the infection was only 15% compared to a reinfection rate of 60% in patients treated with econazole alone. A number of studies have investigated the benefits of *Echinacea* in reducing the number of respiratory tract infections, the severity of symptoms and also the duration of illness with positive results. The consensus view is that *Echinacea* preparations are of value as an adjuvant therapy and for the prophylaxis of recurring infections of the upper respiratory tract and of the urinary tract. Clinical results also support its use in the external treatment of abscesses, boils, chronic skin ulcers and wounds that are slow to heal. Regulatory authorities suggest that *Echinacea* should be used for a maximum of 8 weeks, and it is not recommended in progressive systemic and autoimmune diseases such as tuberculosis, multipe sclerosis, AIDS or HIV infections. In the latter case this is due to a reported reduction in T-cells in an *in vitro* test, but some herbalists believe that this restriction is not justified by the benefits they have observed when AIDS patients have used *Echinacea.*

L04 Immunosuppressive agents

The therapeutic interest in this category has above all been concentrated on the problem of preventing the rejection of an allograft in transplantation surgery.

The pioneers were scientists at Sandoz research laboratories, who isolated the first potent immunosuppressant, cyclosporin, previously known as cyclosporin A, which is a cyclic polypeptide, consisting of 11 aminoacids with a molecular weight of 1202.

In 1970 a research worker from Sandoz collected some soil samples in Hardangervidda, Norway. Two new strains of fungi were isolated; one of them, *Tolypodium inflata* Gams, produces cyclosporin via an industrial fermentation process.

Studies in the 1970s showed that cyclosporin inhibits humoral immunoreactions, and that it had a selective effect on T-cell dependent immunoreactions and that its effect was reversible. Cyclosporin is considered to interfere with the process for primary T-cell activation. In this way the formation of T-effector cells, cytotoxic T-lymphocytes or killer lymphocytes, which have the dominant function in cell mediated immune reactions like rejecting an allograft in transplantation surgery and delayed hypersensitivity reaction, is prevented.

At the present time, the efficacy and the clinical benefit of cyclosporin therapy has been conclusively demonstrated for severely affected patients in four diseases: autoimmune uveitis, psoriasis, idiopathic nephrotic syndrome and rheumatoid arthritis, as well as in transplant patients.

The immunosuppressive agents are beginning to be used in other immunological conditions, such as psoriasis. A drug from Central America has been introduced in Europe: a fern, growing in Central America, *Polypodium leucotomus*, is called locally 'Calaguala'. Its roots and leaves are administered (as an extract) orally in the therapy of psoriasis.

M Musculo-skeletal system

M01 Antiinflammatory and antirheumatic products

M01 A Antiinflammatory/ antirheumatic products, non-steroids

POPLAR *Populus tremuloides*, Michx., family *Salicaceae*, gives the drug *Populi tremula cortex*, poplar bark.

The bark contains salicin (2–4%), salicortin, salireposide and benzoate derivatives: populin (salicin-6-benzoate), tremuloidin (salicin-2-benzoate) and tremulacin (salicortin-2-benzoate). Other constituents are tannins, triterpenes, amyrin and β-amyrin.

Poplar has antirheumatic and anti-inflammatory, antiseptic, astringent properties and has been used for muscular rheumatism, cystitis, common cold and specifically rheumatoid arthritis. The effect is due to the salicylic acid liberated from the salicin-type glycosides by hydrolysis to the aglycone-salicyl alcohol which is then oxidised in the liver to salicylic acid.

Populi nigrae gemmae and *Populi tremulae gemmae* are the buds of these two *Populus* species. They are used as expectorants and circulatory stimulants for the treatment of upper respiratory tract infections and rheumatic conditions.

WILLOW *Salicis albae cortex* is obtained from *Salix alba* L. The bark can also be obtained from *Salix fragilis* L., *Salix pentandra* L. and *Salix purpurea* L., belonging to the family *Salicaceae.* The constituents are qualitatively similar involving various phenolic glycosides, including salicin, salicortin, saliresoside, and their acetylated derivatives. Salicylate content varies between species: 0.5% (calculated as salicin) in *S. alba*; 1–10% in *S. fragilis*; 3–9% in *S. purpurea*. Pharmacopoeial quality bark contains not less than 1% of total salicin.

Indications are: feverish conditions, symptomatic treatment of mild rheumatic complaints, and relief of pains including mild headache.

The onset of the antiinflammatory effect of willow bark constituents is delayed in comparison with that saligenin (salicyl alcohol), sodium salicylate and acetylsalicylic acid, indicating that metabolites of willow constituents may be the active principles.

Salicin (Salix) — CH_2OH, O–Glu — Intestine → Salicylalcohol + Glucose — CH_2OH, OH — Liver → $COOR_1$, OR_2

Salicylic acid: $R_1 = R_2 = H$
Methyl salicylate $R_1 = CH_3$, $R_2 = H$

MEADOWSWEET *Spiraeae herba*, *Spiraeae flos*, is obtained from *Filipendula ulmaria* L., belonging to the family *Rosaceae*.

The chemistry is characterised by a number of phenolic constituents, including flavonoids (Hyperoside, spireoside, kaempferol glycoside), salicylates (salicylaldehyde, gaultherin, methylsalicylate, salicin, salicylic acid), tannins (water-soluble, hydrolysable types, also some catechols). The volatile oil contains salicylates, benzaldehyde, ethylbenzoate, heliotropin, phenylacetate and vanillin.

Documented scientific evidence justifies some of the antirheumatic, antiseptic and astringent properties, but no clinical data are available. An antiulcer activity has been found in aqueous extracts, with the greatest activity in flowers. Some bacteriostatic activity is also found.

DEVIL'S CLAW The drug *Harpagophyti radix* is the secondary root tubers of *Harpagophytum procumbens*, (Burch), D.C., family *Pedaliaceae*, growing in the Savanna regions of Kalahari (Southern Africa).

The characteristic constituents are iridoid glycosides: harpagoside, harpagide and procumbide. It contains not less than 2.2% of iridoid glycosides calculated as harpagoside or not less than 1.0% harpagoside.

Clinical studies with Devil's claw preparations have shown positive benefits in patients with arthroses and pain of rheumatic origin. Pharmacological studies have been less clear cut with results from semi-chronic models being more convincing than those in the standard acute experiments, e.g. the carrageenan-induced oedema test. The effect of gastric juice on the activity is controversial since the clinical results appear to contradict the suggested need for enteric protection of the preparation arising from pharmacological test results.

TURMERIC The drug *Curcumae rhizoma* is the dried, processed rhizome of *Curcuma longa* L., syn. *C. xanthorriza* Roxb., belonging to the family *Zingiberaceae*. The plant is cultivated throughout the tropics.

The drug contains 6% of a pale yellow to yellow-orange volatile oil composed of a number of monoterpenes and sesquiterpenes, including zingiberene, curcumene, α - and β-tumerone among others.

The colouring principles (5%) are curcuminoids, 50–60% of which are a mixture of curcumin and two of its derivatives.

The principal use of the drug today is in treatment of acid, flatulent or atonic dyspepsic ulcers, and pain and inflammation due to rheumatoid arthritis. The antiinflammatory activity of *Curcumae radix* has been demonstrated in animal models, its effectiveness being similar to hydrocortisone acetate and indomethacin. Curcumin and its derivatives are the active antiinflammatory constituents, and its activity appears to be mediated through the inhibition of the enzymes trypsin and hyaluronidase. Clinical studies (randomised, double-blind) have verified the anti-inflammatory action and effectiveness in the treatment of dyspepsia.

The choleretic and cholecystokinetic activity is described in section A05AX.

GINGER (*Zingiber officinale Roscoe*) has also shown beneficial effects in rheumatoid conditions.

M02 Topical products for joint and muscular pain

Counter-irritant or rubefacient drugs are all founded on the same principle. In an ointment, liniment or plaster there are substances which exert a strong vasodilating action on the circulation in the skin, followed by an intensive, persistent feeling of warmth and also a vasodilatation in deeper-situated tissues. These effects provoke what is wanted therapeutically: a local symptomatic relief in joint and muscular pain. By combination of active components, the effects of which have a different onset of action, a long and strong effect might be obtained. The indications for all such remedies are joint and muscular pain, both acute and chronic, lumbago, neuralgia, ischias, muscular spasms.

CAPSICUM The drug *Capsici frutescens fructus* is the dried fruit of *Capsicum frutescens* L., family *Solanaceae*, whereas paprika and Spanish pepper are various cultivars of *Capsicum annuum*. The sharp, burning taste is given by capsaicin (major component, 49%) and derivatives of capsaicin. The content of capsaicin is much higher in *C. frutescens* than in *C. annuum*, while the variety which is used as a vegetable has almost no capsaicin.

Capsaicin is the most important of a number of vanillyl fatty acid amides from capsicum, referred to as capsaicinoids, and is largely responsible for the hot pungent effect of chilies. Capsaicin causes the release of substance P from nerve terminals and as a result the intense burning pain and subsequent loss of pain sensation associated with highly spiced food. This loss of pain sensation is exploited clinically in cases of post herpetic neuralgia and also in peripheral neuropathy due to diabetes, in stump pain in amputees and in trigeminal neuralgia. After *Herpes zoster* (shingles) infections, treatment with capsaicin externally four times per day for several days may give relief from pain after the initial skin reaction subsides. Capsicum and its oleoresin may also be used as a counterirritant in rheumatism, arthritis and lumbago, although capsaicin levels in the oleo-resin depend on the variety of capsicum used and its maturity.

CAMPHOR The drug from the essential oil, obtained by steam distillation of the wood from *Cinnamomum camphora* (L.), J. Presl, belonging to the family *Lauraceae*, a high tree that is cultivated mainly in Taiwan. The production of natural camphor (optically active) has been replaced by synthetic camphor (racemate), which is synthesized from α-pinene (from turpentine oil).

Camphor is a good example of the importance of the mode of application for the cutaneous effect. If camphor is massaged intensely in the skin the result will be an erythema, but if it is spread out carefully and lightly, a cooling effect is obtained, similar to the effect of menthol. Furthermore, camphor has a local anaesthetic effect.

In external use Camphor exerts a hyperaemic action; by inhalation it is bronchosecretolytic and in oral administration it acts as central stimulant.

EUCALYPTUS OIL *Eucalypti aetheroleum* is obtained by steam distillation of fresh leaves (in practice whole twigs) of *Eucalyptus globulus* Labill. and other *Eucalyptus* species, such as *E. fructicetorum* F. von Müller (=*E. polybractea* R.F. Baker), family *Myrtaceae*. The genus *Eucalyptus* belongs to the Australian flora, but now has a global distribution by extensive cultivation, because of rapid growth and absorption of water from swampy regions.

The main constituent is eucalyptol (1,8-cineole) 70–85%, the rest being various monoterpenes.

Expectorant and antibacterial activities have been reported for eucalyptus oil and for eucalyptol, and eucalyptus oil is taken orally for catarrh or used as an inhalation. (See RO5C.)

ROSEMARY The spice rosemary, *Rosmarini folium*, is the dried leaves of *Rosmarinus officinalis* L, family *Lamiaceae*, which is an evergreen shrub in the Mediterranean region. The leaves contain volatile oil, *Rosmarini aetheroleum* (1-2.5%) with α- and β-pinene, cineole, borneol and camphor (10-20% of the oil).

Antibacterial and antifungal activities *in vitro* have been reported for the oil, as well as spasmolytic action in smooth muscle.

Traditionally rosemary is indicated for flatulent dyspepsia, headache, and topically for myalgia, sciatica and intercostal neuralgia.

LAVENDER Lavender oil, *Lavendulae aetheroleum*, is distilled from the fresh flowering tops of *Lavandula angustifolia* Mill.(= *L. officinalis Caix et Mill.*), family *Lamiaceae*.

The essential oil contains 30-60% esters (mainly linalylacetate), limonene and cineole. The essential oil is a common constituent in rubefacient preparations. Most lavender oil is used in the perfumery industry and aromatherapy.

MO3 Muscle relaxants

MO3 A muscle relaxants peripherally acting agents

CURARE ALKALOIDS The term 'curare' is a generic one applied to various South American arrow-poisons used by Indians for hunting. (Fig. 17) These extracts are made from a number of different plants, particularly members of the *Menispermaceae* and *Loganiaceae* families.

Curares from the upper Amazon (Brazil and Peru) seem to be mainly Menispermaceous in origin, while those from British Guiana, Venezuela and Colombia owe much of their activity to species of the genus Strychnos (family *Loganiaceae*), especially *Strychnos toxifera*, *S. castelnaeana* and *S. guianensis*.

Fig. 17. The Indians in Eastern Peru (Amazonas region) use curare on the arrows in their blow pipes.

Here the curare-alkaloids are divided according to botanical origin into:

MENISPERMACEAE-CURARE This type of curare is prepared from the bark of *Chondrodendron tomentosum* Ruiz and Pavon and *C. platophyllum* Mers and other *Chondrodendron* species, family *Menispermaceae*, by extraction with boiling water followed by evaporation to a syrup. (Fig. 18) The most important active principle is tubocurarine, which is a dimeric benzylisoquinoline alkaloid with one quaternary nitrogen.

The muscle-relaxing effect is obtained because the alkaloid increases the threshold in the motor endplate for the acetylcholine, liberated at the nerve transmission. The effect of tubocurarine can therefore be

Fig. 18. *Chondrodendron platophyllum* Mers

counteracted or abolished by cholinesterase inhibitors like physostigmine or synstigmine. The duration of action of tubocurarine is relatively long.

LOGANIACEAE - CURARE To prepare this type of curare, the bark of *Strychnos toxifera* Rob. Schomb. ex. Lindl., *S. guianensis* (Aublet) C.Martins, *S. castelnaeana* Wedd. and closely related species of the family *Loganiaceae* is used. Among the many alkaloids that occur in these barks, toxiferine-I should be mentioned, a dimeric indole alkaloid (with strychnine-skeleton) which has two quaternary nitrogens.

Toxiferine is not used clinically as such, but a semisynthetic derivative: N,N-diallyl-bis-nor-toxiferin, Alkuron, where the methyl-groups at the two quartenary nitrogens are substituted by allyl-groups, is used.

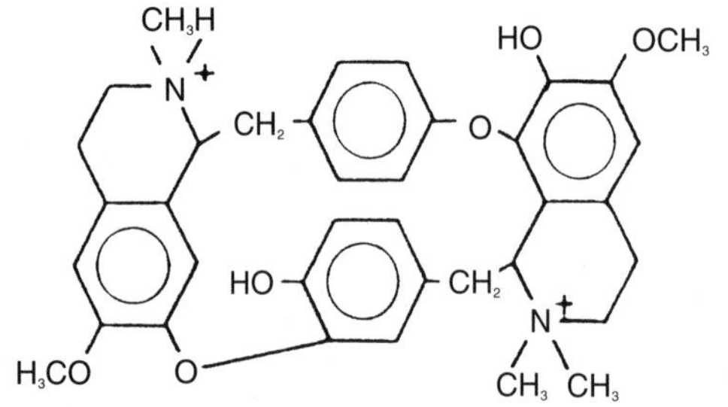

(+) –Tubocurarine salt

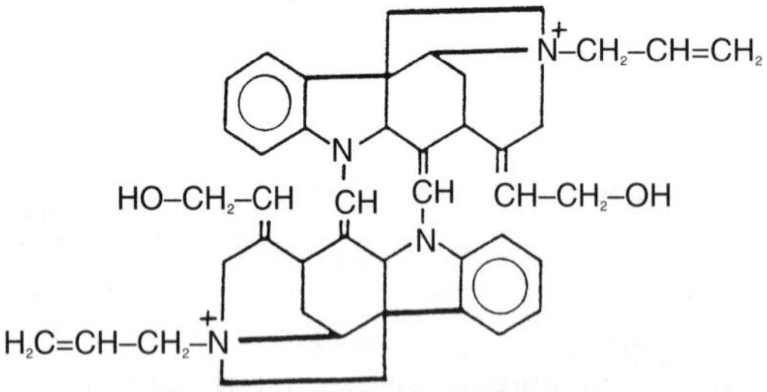

Alloferin R (Alkuron)
N,N′ Diallyl-nor-toxiferin

Alkuron has the same mode of action as tubocurarine, but has 1.5-2 times its potency, the duration of action being somewhat shorter. The latency is about 30 seconds after intravenous injection and full muscle relaxation is obtained after 2-3 minutes. By individual adaptation of doses and intervals of administration the intensity and duration of the effect can be controlled.

MO4 Antigout preparations

COLCHICINE This alkaloid is dealt with in the section LO1C C among the cytostatics. Here it is used for the treatment of acute attacks of gout in a dose of 0.5 mg/hour until the pains disappear, the maximal daily dose being 4-5 mg. It can also be used for differential diagnosis: if the pains disappear after colchicine treatment, it is a case of *arthritis urica*, otherwise not.

Colchicine suppresses chemotactic properties in leukocytes and inhibits the inflammation reaction. Colchicine is easily absorbed from the intestinal tract and 40% is excreted in the urine within 40 hours. There is risk of accumulation on repeated administration.

MO5A B Antiinflammatory enzymes A variety of animal and plant proteases are used to reduce inflammation and swellings in wounds and abscesses after minor dental surgery, for example. The plant enzymes include papain, bromelains and ficin. These and animal enzymes such as trypsin and chymotrypsin (pancreatic enzymes) are used as debriding agents in burn tissue, pus etc.

N Nervous system
NO1 Anaesthetics
NO1B Local Anaesthetics

COCA LEAVES, *Cocae folium* The drug is the dried leaf of *Erythroxylon coca*. Lam. or E. *novogranatense* (Morris) Hierow,

Fig. 19. *Erythroxylon coea* Lam. Cultivation in Tingo Maria, Peru

Fig. 20. Sale of coca leaves in the market in Pisac, Peru

family *Erythroxylaceae.* The coca bush is cultivated in the valleys of the Andes, at about 700–1,000 m above sea level from Colombia to Bolivia. (Fig. 19)

The most important ingredient of coca leaves is the alkaloid cocaine, which has local anaesthetic and central stimulant effects. Which effect should be considered the desired, intended effect and which should be considered a side-effect depends in this case entirely on the intended use.

Cocaine

The introduction of cocaine as a local anaesthetic was of great importance and made possible ophthalmological operations that previously were not feasible. For instance Koller in Vienna introduced cocaine in operative ophthalmology. The chemist Willstätter elucidated the structure of cocaine in 1898 and it has been the model for the synthetic local anaesthetics for over nine decades (procaine, lidocaine, etc.).

For the coca-chewing Indians in South America, on the other hand, the central stimulant effect is the wanted effect, whereas the local anaesthetic action is an undesired side-effect. The dried leaves (which are purchased on the local market, (see Fig. 20) are mixed with alkali (burned bones from llamas, which is called Lliptar); the mixture is kept between the teeth and cheek, without direct chewing, and the saliva is swallowed. Coca chewing gives rise to less dependence and social problems than the use of isolated cocaine in the form of inhaled cocaine hydrochloride or the smoked free-base alkaloid called ‘Crack’ or ‘Free-base’.

NO2 Analgesics

NO2A Opioids

OPIUM The drug *Opium* is the latex from unripe capsules of *Papaver somniferum* L., family *Papaveraceae,* dried in the air. The opium poppy, *Papaver somniferum* L., has been known for several thousands of years as a pain-killing and narcotic remedy. (Fig. 21) It is an annual plant, 1–1.5 m in height with blue-green leaves and 4 white, violet or purple petals, which fall off before the egg-shaped capsule is ripe. The fixed oil in the seeds is used for the production of margarine. For the production of opium the

Fig. 21. *Papaver somniferum* L. Horizontal incisions in unripe capsules with fresh latex in the form of white droplets.

poppy is cultivated in Tasmania, Turkey (certain areas of Anatolia) and India, but also on a small scale in former Jugoslavia, Russia and China. In these countries the cultivation is controlled by the International Narcotics Control Board of the United Nations. Illegal production of opium is found in Turkey, India, Afghanistan and the countries in the 'Golden Triangle', i.e. the borders between Laos, Thailand and Burma. However, nowadays very little is grown in Thailand, where the opium poppy is substituted by other crops (King's Highland Project). The unripe capsule of the opium poppy has an anastomosing system of latex vessels, containing a yellow-white latex. For opium production, thin incisions are made in the capsule; the latex appears as white drops, which rapidly turn brown and become semi-solid in consistency. The incision or incisions are made in the capsule wall during the afternoon. The following day the dried latex is scraped off (crude opium, dark brown-black colour) giving about 20–50 mg per capsule, corresponding to 2–5 mg morphine. After additional drying in the air, the crude opium is formed into lumps, which are sent to opium factories (in Turkey: Istanbul and Izmir; in India: Ghazipur) where they are further worked up and packed for export. The annual total official production of crude opium is of the order of 2,000 tons.

Crude opium, *Opium crudum*, Pharm. Eur. has a morphine content of 12–16% whereas the pharmacopoeias require 10%. The alkaloids are bound to meconic acid, lactic acid, fumaric acid and sulphuric acid. Opium has a very characteristic smell and can be identified microscopically by the occurrence of characteristic fragments from the wall of the capsule.

Crude opium is used as the starting material for the extraction of morphine (of which 80% is methylated to codeine), codeine and noscapine. Opium, according to the pharmacopoeia, is used for the preparation of opium tincture, which is an excellent remedy for the treatment of tropical diarrhoea.

The following table shows the contents of various alkaloids in crude opium:

	Opium alkaloid	% content in crude opium
Morphinan-type	Morphine	3.0–23.0
	Codeine	0.3–3.0
	Thebaine	0.3–1.0
Benzyliso-Quinoline-type	Papaverine	0.8–1.5
	Noscapine (=narcotine)	2.0–12.0
	Narceine	0.1–0.2

OPIUM FOR SMOKING (*schandy, Chandu*) Crude opium is extracted repeatedly with water, then fumigated and fermented for several months to get a product with a fine, peculiar aroma, which is smoked in special pipes. Daily dose is 5–15 g, corresponding to 0.5–1.5 g of morphine.

MORPHINE EXTRACTION FROM POPPY STRAW The production of opium is, as is evident from the description above, a method that is very labour intensive. Therefore methods have been developed to extract morphine directly from the dried capsule wall after the fixed oil has been harvested without the necessity for lancing for opium production.

In Eastern Europe, e.g. Hungary and Romania, the opium poppy is cultivated for the seeds and poppy straw is a useful by-product.

The antitussive effect of codeine and noscapine is dealt with in RO5, the spasmolytic effect of papaverine is dealt with in AO3A and the constipating effect of opium tincture is mentioned in AO7.

MORPHINE When morphine is used as an analgesic, it reduces pain and causes sedation. The pain is registered, but not experienced as unpleasant or threatening. The perception of time and memories are disturbed. Other central effects include depression of respiration, which usually is without clinical importance, but in the case of massive overdosage is the cause of death, and a vagal-stimulation, which gives a pronounced miosis, indisposition and vomiting. As specific antidotes in morphine overdose, naloxone and nalorphine are used. Since these two substances are short-acting, they have to be administered repeatedly.

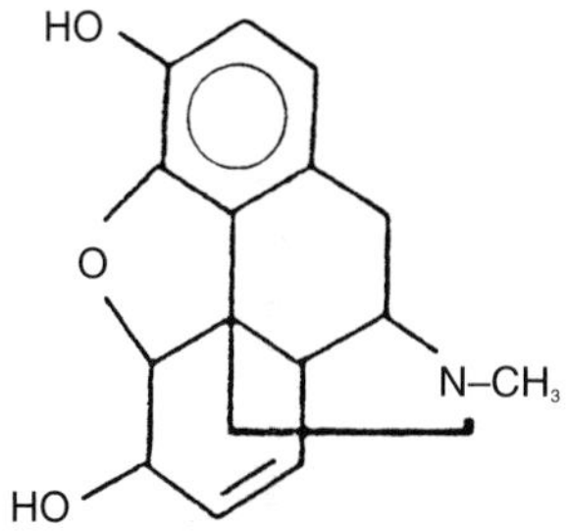

HO O N–CH_3 HO

Morphine

As a classical pre-medicine before operations morphine + scopolamine: 10 mg + 0.4 mg is used. Morphine is used to make diacetylmorphine (heroin) and codeine (methyl morphine).

NO2B Other Analgesics and Antipyretics

ROMAN CHAMOMILE, *Chamomillae romanae flos*, is the flowers of *Chamaemelum nobile* (L.) All. (*Anthemis nobilis* (L.)), family *Asteraceae*. Contains at least 0.7% essential oil. This plant is mainly used against dysmennorhea, often as an ingredient in a herbal tea.

ACONITE, *Aconiti radix*, the tuberous root of *Aconitum napellus* (L.), family *Ranunculaceae*, is a classical analgesic drug, which was formally used. As a homoeopathic remedy it is used in the dilution D6 (1:1 million) and thus without any effect according to the concepts of conventional medicine.

NO2C Antimigraine preparations

NO2C A A Ergot alkaloids Ergot, *Secale cornutum*, is the sclerotium (the hibernating part) of the fungus *Claviceps purpurea* (Fries) Tul., family *Hypocraeaceae*, Class Ascomycetes, which is a parasite on rye, *Secale cereale* (L.), family *Poaceae*. (Fig. 22)

LIFE CYCLE The ergot undergoes a developing cycle, starting with the germination of the sclerotium in the springtime, developing ascospores. These are blown by the wind to the flowering rye (or another grass), where they penetrate the ovary (gynoecium) and germinate and also produce another type of spore, conidiospores, which are spread by insects to other rye plants, and transform the gynoecium to a lengthened, bent, dark brown or black, solid mass of hypheae, which grow bigger than the normal fruit of rye: 1–5 mm in diameter. In the autumn the selerotium falls to the earth and hibernates, grows the next spring and thus the circle is completed.

Fig. 22. Ergot, *Secale cornutum*, is the sclerotium (the resting stage) of the fungus *Claviceps purpurea* (Fries) Tul., a parasite on rye, *Secale cereale* L.

HISTORY: In earlier days the sclerotia were considered as a welcome contribution, which increased the bulk of of rye-meal. At that time one did not understand the risks of ergot poisoning; and therefore 'epidemics' occurred after the consuming of ergot-contaminated rye-meal, even up to the twentieth century. A characteristic symptom was gangrene in the feet and hands owing to the contraction in the peripheral blood vessels. An early symptom, called St. Anthony's Fire, was a burning feeling in the extremities.

CHEMISTRY OF THE ERGOT-ALKALOIDS The ergot alkaloids have an ergoline skeleton, which forms a part of the common ingredient D-lysergic acid, which has two asymmetric C-atoms in positions C-5

D-Lysergic acid

Ergoline

Ergometrine

and C-8. The hydrogen at the C-5 position is always in the β-position in all alkaloids found in nature. The pharmacologically active alkaloids, which are derivatives of D-lysergic acid, have another hydrogen in the C-8 in the α-position. D-lysergic acid derivatives can easily be transformed to the inactive D-isolysergic preparations, where names are given the ending-inine, e.g. ergometrinine, ergotaminine.

D-lysergic-acid-alkaloids can be divided into two groups:

1. Water-soluble lysergic acid amides: ergometrine (ergobasine) and the semi-synthetic methergine.
2. Cyclopeptide-group with two sub-groups:
   a) the ergotoxine group: consisting of ergocristine, ergocryptine and ergocornine.
   b) the ergotamine group.

Among the lysergic acid amides, the alkaloid ergometrine (ergobasine) and its methyl-derivatives (Methergine) are the most important. They have a pronounced contraction effect on the pregnant uterus and have been discussed in section GO2A.

The ergotoxine group consists mainly of hydrogenated alkaloids in the preparation Co-Dergorcine (see CO4A).

ERGOTAMINE Ergotamine has a sustained contractile effect on the uterus and can be administered to prevent bleeding after delivery.

In the prodromal stage of a migraine attack the physiological blood circulation in the capillary bed in the brain is substituted by opening of arteriovenous anastomoses (shunts) through which there will be a diminished circulation in the brain and diminished oxygen supply. Migraine pain is explained by the cortical vasodilation of these shunts. In migraine the neurotransmitters serotonin and tryptamine play an important role. It is known that serotonin-stimulation of α-receptors in the cell walls of the cerebral vessels gives a vasoconstriction of the arteriovenous shunts and a relief of the pain of migraine. When serotonin-content drops there is a passive dilation of the extracranial blood vessels, therefore ergotamine, the therapeutic effect of which is obtained by constriction of hypotonic arteries via stimulation of adrenergic alpha-receptors, should be administered as early as possible during a migraine attack.

The central emetic effect is an unwanted side-effect.

Ergotamine is a serotonin antagonist, which gives a vasoconstriction of intra- and extra-cranial arteries, above all the external carotid artery, and furthermore intensifies the effects of noradrenaline.

Two other ergot alkaloids are used in the prophylaxis of migraine, namely

DIHYDROERGOTAMINE This alkaloid is obtained by partial hydrogenation of ergotamine. It is used in the treatment of migraine as a relatively weak vasoconstrictor.

METHYSERGIDE Methysergide is a semisynthetic lysergic acid amide alkaloid. It is obtained from N-methyl-lysergic acid by reaction with butylamine. Methysergide is a strong antagonist of 5-HT_2-receptors with special affinity to 5-HT_1 receptors. It is mainly used for the prophylaxis and treatment of migraine.

PRODUCTION OF ERGOT ALKALOIDS The simplest method is collection of the wild drug, which is performed in Portugal and Spain. Cultivation of ergot on rye is another method. This is performed in India and Nepal, where there are races which mainly produce either ergometrine or ergotamine or ergotoxine. The newest method used to produce alkaloids is cultivation of the fungus in tanks analogously to the production of antibiotics. Dr Tonolo at Instituto Superiore di Sanità in Rome succeeded in producing lysergic acid amides from the fungus *Claviceps paspali*, which he found to be a parasite on the grass *Paspalum districtum* (L.) in a coastal meadow south of Rome. The lysergic acid amides which are produced by this fungus can then chemically be transformed to alkaloids, which are used therapeutically.

NO2C X Other antimigraine preparations

FEVERFEW, *Tanacetum parthenium* (L) Schultz Bip. syn. *Chrysanthemum parthenium* (L.) Bernh., family *Asteraceae*. The main active ingredients of Feverfew leaves appear to be the sesquiterpene lactones, including parthenolide, which have an antiphlogistic action. Extracts of Feverfew inhibit human blood platelets and polymorphonuclear leucocytes. The effectiveness of Feverfew in migraine prophylaxis has been demonstrated by clinical studies. The migraine prophylactic action is thought to be due to inhibition of the release of serotonin from platelets, possible via neutralisation of sulphydryl groups on specific enzymes that are fundamental to platelet aggregation and secretion. After 4 months of treatment the frequency and intensity of migraine attacks including nausea and vomiting were diminished compared to placebo, but the duration of each attack was unchanged. The sesquiterpene lactones can also cause contact dermatitis and particularly aphthous or mouth ulcers. Patients who are allergic to other members of the *Asteraceae* may be at risk.

NO4 Anti-parkinson drugs

Bromocriptine (2-brom-ergocryptine) (see GO2C)

NO5 Psycholeptics

This group is divided into three therapeutic subgroups:

NO5A Antipsychotics
NO5B Anxiolytics
NO5C Hypnotics and sedatives

NO5A Antipsychotics

This group comprises drugs with antipsychotic actions (i.e. neuroleptics).

RESERPINE is classified in CO2, Antihypertensives.

NO5B Anxiolytics

This group comprises preparations used in the treatment of neuroses and psychosomatic disorders associated with anxiety and tension.

NO5B X Other anxiolytics

ST. JOHN'S WORT, *Hyperici herba*, is the dried aerial parts of *Hypericum perforatum* L., family *Hypericaceae. Hypericum* contains flavonoids, xanthones, naphthodianthrones such as hypericin and pseudohypericin and phloroglucinols, e.g. hyperforin. Hypericin has anti-HIV activity and is responsible for the photosensitivity reaction which has been documented for St. John's Wort in cattle and sheep. It is believed that levels of hypericin in commercial products are unlikely to cause such reactions in humans but excessive use and exposure to sunlight after use should be avoided. *Hypericum* is a highly popular herb because of its documented clinical effectiveness as an antidepressant. A meta-analysis of 23 trials of which 15 were against placebo and eight compared the plant extract with standard antidepressants showed that *Hypericum* was superior to placebo in alleviating the symptoms of depression and that it was as effective as some synthetic antidepressants. It now appears that no one group of compounds explains the effects of *Hypericum*. Hypericin, the xanthones, the flavonoids and hyperforin affect various enzymes and transmitter systems, including monoamine oxidase A+B, benzodiazepine, GABA and NMDA receptors.

There is now evidence that *Hypericum* extracts induce cytochrom P450 and because of this effect patients using cyclosporin, digoxin, SSRI (synthetic anti-depressants), warfarin, inclinavir or oral contraceptives must be counselled not to use *Hypericum* or if they are already doing so, not to stop use abruptly.

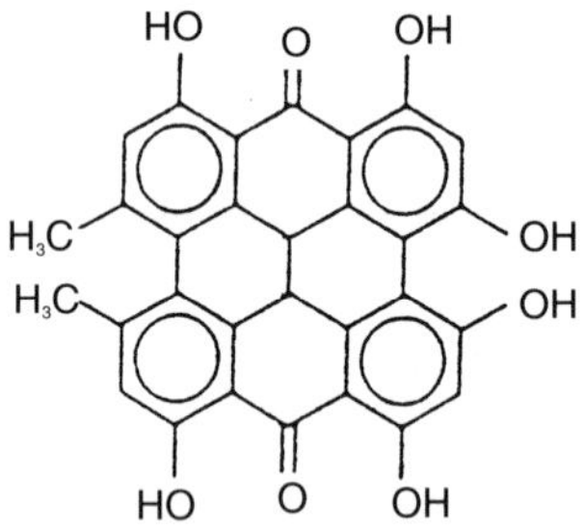

Hypericin
(Hypericum perforatum)

NO5C Hypnotics and sedatives

VALERIAN ROOT, *Valerianae radix* Valerian root consists of the underground organs of *Valeriana officinalis* L.s.L. (*Valerianaceae*) (Fig. 23) including the rhizome, roots and stolons, harvested during

Fig. 23. *Valeriana officinalis* L

early autumn of the second year of growth, washed and immediately dried at a temperature below 40°. It contains not less than 0.5% V/m of essential oil. Valerian is a perennial herb that grows in the whole of Europe, cultivated in Central and Eastern Europe and at USA. The fresh root has no smell, but when dried it has a spice-like odour, attributed to bornyl isovalerianate and free isovalerianic acid. Valerian root has several types of active components:

a) the essential oil (0.5–1.5%) containing monoterpenes (such as bornyl esters, camphene and pinenes), sesquiterpenes (including valerenal and valeranone).
b) less volatile sesquiterpene acids (valerenic acid, and derivatives).
c) gamma-aminobutyric acid (GABA) glutamine and arginine in relatively high concentrations.
d) Valepotriates (*valeriana-epoxy-tri esters*) are found in fresh and carefully dried drug (0.5–2%). They are unstable triesters of alcohols with an iridoid nucleus, which also has an epoxide nucleus. The acids are acetic acid, isovaleric acid, acetoxy-isovaleric acid. The relative content of valepotriates is different in the root of the three species that at present are used therapeutically: in *V. officinalis* it is the main component (80–90%); *V. wallichii* (Pakistan, India) has two chemical races, a didrovaltrate race and an acevaltrate race; *V. edulis*, a Mexican species, contains isovaltrate as the dominant constituent. The valepotriates are unstable and are unlikely to be present in finished products, especially those based on aqueous extracts. Similarly, baldrinals, the decomposition products of valepotriates, are generally not detected in valerian root preparations. The stability of valepotriate preparations is a galenical problem: dragees are probably the best.
e) Lignans with an affinity for 5 - HTIA receptors.

Sedative properties have been documented for valerian and have been attributed to both the volatile oil and the valepotriate fraction. Screening of the volatile oil components for sedative activity concluded that valerenal and valerenic acid were the most active compounds.

Valerenic acid inhibits the enzyme system responsible for the central catabolism of GABA. Increased concentrations of GABA give a decrease in CNS activity, which is involved in the sedative action of valerenic acid.

COOH

Valerenic acid

CNS depressant activity after injection of valepotriates and their degradation products has been documented. A specific valepotriate fraction, Vpt_2, shows tranquilising, central myorelaxant, anticonvulsant, coronarydilating and antiarrythmic action in animals.

At least 11 clinical trials with Valerian have been conducted; most showed improvements in sleep quality and sleep latency (time taken to fall asleep). These effects were more noticeable in patients who were rated as poor sleepers than in healthy subjects. In one study, for example, changes were noted after taking Valerian and in another over 11,000 patients paticipated in an open study for 10 days. After taking a Valerian extract, there was a 72% success ratio in cases of difficulty in falling asleep, 76% in cases of discontinuous sleep and 72% in patients who were restless and tense. Among the benefits of using Valerian is that it is very safe in terms of overdosage (LD_{50} for an ethanolic extract is reported as 3.3 g/kg of body weight in mice); there is no risk of dependence as there is with benzodiazepines, nor is there any evidence of synergistic effects with ethanol: much of the evidence suggests that using Valerian does not affect a person's ability to drive or use machinery but given that one study did show such an impairment, it is preferable to be cautious and advise patients not to drive after using the drug. There is no doubt that Valerian is a valuable and safe alternative to benzodiazepines in cases of mild to moderate stress and anxiety presenting as tenseness, restlessness and irritability and also in cases of mild to moderate insomnia where there is difficulty in falling asleep.

HOPS, *Lupuli strobilus* Hops consist of the dried inflorescences of *Humulus lupulus* L., family *Cannabaceae*, collected from the female plants. It contains not less than 0.3% (V/m) of essential oil.

Glandular trichomes can be separated from the strobiles by sifting and this material is known as lupulin. Hops contain bitter principles (15–25%) consisting of two types of resin, hard and soft, which are important in beer production.

The lipophilic soft resin consists of a series of phloroglucinol derivatives known as α-acids (humulone) and β-acids (lupulone). Both α- and β-acids are gradually denatured during storage. The hard resin contains both a hydrophilic fraction (δ-resin) and a lipophilic fraction (γ-resin).

The volatile oil contains mainly myrcene, humulene and caryophyllene. Up to 0.15% of 2-methyl-3-buten-2-ol can be detected in the oil after storage of Hops for 2 years through metabolism of humolones and lupulones, but only traces are found in fresh material. This substance, which structurally and pharmacologically is similar to methylpentynol, is a synthetic sedative-hypnotic drug.

The therapeutic indications are tenseness, restlessness and difficulties in falling asleep.

The sedative effect of Hops is undisputed in empirical medicine but the basis of the action is not yet fully understood.

PASSIONFLOWER, *Passiflorae herba* Passionflower consists of the dried aerial parts of *Passiflora incarnata* L., family *Passifloraceae*, collected during the flowering and fruiting period. This plant is a climber from the Carribean region, and is now cultivated in the whole tropical zone. It contains flavonoids, mainly C-glycosides of apigenin and luteolin with considerable variation in qualitative and quantitative composition.

The therapeutic indications are tenseness, restlessness, irritability with difficulty in falling aseep, nervous stress, and anxiety.

The available pharmacological studies generally support, with some conflicting results, the empirically acknowledged sedative and anxiolytic effects of Passiflora, but it is not yet clear which constituents are responsible for these actions. One possibility which has received attention is related to the ability of flavonoids to bind to benzodiazepine receptors. The acute toxicity is very low.

NO6 Psychoanaleptics

NO6B Psychostimulants

NO6B C Xanthine derivatives The drugs included in the following table all

The occurrence of xanthine-derivatives in some drugs.			
Drug	Caffeine	Theophylline	Theobromine
Coffee beans (roasted)	0.3-2.5	trace	trace
Tea leaf	2.5-5.0	0.02-0.04	appr 0.05
Cocoa seed	0.2-0.3	trace	1.0-2.0
Cola seed	0.6-3.0	-	0.1
Maté leaf	0.5-1.5	0.05	0.45
Guarana	2.5-5.0	-	-

have varying contents of the xanthine-derivatives theophylline, theobromine and caffeine.

All three xanthine-derivatives occur in the living plants loosely bound to sugar, phenols and tannins, but are liberated during the fermentation and roasting processes involved in the production of xanthine-drugs.

Caffeine is extracted from tea dust and during coffee roasting, caffeine sublimes. Caffeine has above all a centrally stimulant (analeptic) effect on the cerebrum, produces tachycardia and has a diuretic effect.

Theobromine is extracted from the shells of cocoa-seeds. Theobromine has no centrally stimulant action but instead it is a diuretic and it has coronary dilating effects. effects.

Theophylline occurs in such a low concentration in all xanthine drugs that extraction is not profitable. Theophylline can be produced by demethylation of caffeine or by total synthesis. Theophylline has a spasmolytic effect on smooth muscles, which is apparent in a pronounced bronchodilating action, hence it is used in asthma either on its own or combined with ethylenediamine to produce aminophylline. Furthermore, theophylline has a diuretic action.

COFFEE, *Coffeae semen* Coffee is the roasted seed from different *Coffea*-species, family *Rubiaceae*, such as *Coffea arabica* L., *Coffea liberica* Bull ex. Hierm. and *Coffea robusta* Lind., which originally gave Mountain coffee, Liberian coffee, and Congo coffee, respectively. Now coffee is cultivated on a global scale with Brazil as the biggest producer followed by Colombia.

The fruit is, when ripe, a red drupe containing white seeds, the seed coat of which has a thin parchment-like membrane, which is called silver membrane. The seed consists mainly of endosperm. The coffee-bean consists of the peeled seed, which is roasted at a temperature of 200-250°C, when the caffeine is partly sublimated, and the aromatic substances are formed.

The unroasted coffee contains 0.3-2.5% caffeine, bound to chlorogenic acid. A cup of normal coffee (appr. 5 g coffee) contains 50-100 mg caffeine, appr. 50 mg 'coffee-oil' and appr. 0.2 g chlorogenic acid.

TEA, *Theae folium* The tea plant is *Camellia sinensis* L., belonging to the family *Theaceae*, from which two kinds of tea are prepared: fermented black tea and unfermented green tea. Tea plantations are found mainly in Sri Lanka, India, China and Japan; nowadays also in Kenya and South America.

The production of black tea starts by withering the leaves for 20-40 hours, depending on temperature and humidity of the air; afterwards the leaves are crushed by rolling, when the polyphenols in the cell juice are oxidised and the phlobotannins are converted into phlobaphenes, a process that is terminated by exposure to the air at a temperature of about 30°C. The oxidation is finished by a rapid drying at 115–120°C and finally sifting.

The oxidised polyphenols give the black colour, its aroma depends on the occurrence of a volatile oil and the tannins of a catechin type give the harsh taste. The caffeine content of a cup of tea is approximately 30–40 mg.

MATÉ, *Matae folium* The drug consists of the dried leaves of *Ilex paraguariensis*, St. Hill. and other caffeine-containing *Ilex* species, belonging to the family *Aquifoliaceae*. This species is an evergreen tree, growing in southern part of Brazil, Paraguay, Uruguay, Northern Argentina. At harvesting the tops of the branches are passed over an open fire, the enzymes are inactivated and the leaves keep their light green colour. Afterwards grinding and storage take place.

Within the area of cultivation, maté is a national beverage, which is sucked from an empty calebass (mate) through a metallic sucker (bombilla). The caffeine content of a cup of Maté tea is approximately 20–30 mg.

COLA NUT, *Colae semen* The cola nut is obtained from the West African trees *Cola acuminata* (P.Beauv.) Shott et Endl. and *Cola nitida* (Vent.) Schott et Endl., family *Sterculiaceae*. Nowadays there are cultivations in Madagascar, India, Fiji, South America and Central America. The drug consists of the dried embryo and its two big cotyledons. *Cola acuminata* gives small cola nuts ('Kolas quarts'), whereas *Cola nitida* gives big two-parted cola nuts ('Kolas demis'). An extract of cola nut is a constituent in Cola beverages, which contain 10–30 mg caffeine per average portion in addition to decocainised extracts of Coca leaves.

COCOA SEED, *Cocoa semen* Cocoa seeds are obtained from *Theobroma cacao* L., belonging to the family *Sterculiaceae*, a tree in Central America which is now cultivated in the whole tropical zone with Ghana as the main producer. After roasting, the seed husk is removed and extracted for theobromine and the rest of the seed, which mainly consists of the cotyledons, is pressed with heat to obtain the Oil of Theobroma (cocoa butter) used for suppositories. The rest, which contains about 20% fat, is powdered to give the drug cocoa.

Plain or bitter chocolate is a mixture of ground cocoa nibs with sucrose, cocoa butter and flavouring. Milk chocolate contains in addition milk powder. A bar of chocolate contains about 6 mg of caffeine and a cup of cocoa about 13 mg.

GUARANA SEEDS, *Guaranae semen* The plant *Paullinia cupana* Kunth, family *Sapindaceae*, grows along the Orinoco River in Venezuela and in the Amazonas region in Brazil, where cultivation is also found.

The drug consists of a hard paste in cylindrical rolls, 10–30 cm long and 2.5–4 cm in diameter. The seeds are collected from wild and/or cultivated plants by members of the Guaranis tribe. The kernels are roughly separated from the shell, broken and made into a paste with water, starch and other substances being frequently added. The paste is then made into suitable shapes and dried in the sun or over fire. The drug has no marked odour but an astringent bitter taste due to tannins. When mixed with water it gives a stimulating drink, which is one of the most caffeine-rich drinks available. In the West it is included in Cola drinks and is also available in tablet and capsule form. One tablet could provide as much caffeine as a cup of coffee.

NO6B E Psychostimulants that generally increase the physical capacity The medicinal plants that are considered to have this effect and which are described here are *Schizandra chinensis, Panax ginseng, Eleutherococcus senticosus, (Acanthopanax senticosus), Rhodiola rosea*, and Pollen.

These plant species (not Pollen), according to the Russian physiologist and pharmacologist Prof. I.I. Brehkman, belong to the group that are called *adaptogens*, i.e. drugs, which increase (or support) the capacity of the organism to attain homeostasis when exposed to stress.

The important properties of adaptogenic substances according to Prof. I.I. Brehkman are:

1. They are harmless and have no side-effects.
2. The adaptogenic effect is general and unspecific, and is not localised to a specific organ. It increases the general capacity of the body to adjust, and therefore increases the resistance of the body.
3. Adaptogenic substances have a normalising action on the functions of the body.

Implicit in this definition of adaptogenesis is that they are not curative, instead they restore an irregular imbalance in the body.

When the strain is of long duration or is continually repeated, the cells 'adapt' to work at a higher level. In this case the cell must be built up to manufacture cell nutrients. A healthy body can manage this adjustment. The problem instead is the higher level of 'waste products', implied in the higher production. An example of this is the so-called 'free radicals'. If these are not neutralised, they harm and destroy the cells. In this process the cells require anti-oxidising substances to neutralise the dangerous radical. Thus, the adaptogenic substances have the following mechanisms of action:

1. They increase the capacity of the cells to produce and use the cell nutrients economically during periods of stress and strain. This is mainly carried out by activating the enzyme glucokinase to form the cell nutrient glucose-6-phosphate.
2. They help the cells to build up 'energy factories' by activating mRNA (messenger) and +RNA (transcription).
3. They act as antioxidants towards the free radicals and their effect on the cell membrane. The membrane is the cell's protective enclosure, and regulates all transports in and out of the cell of nutrients and waste products.

Using modern methods it has been shown that this action is partly an immunostimulatory effect. Besides the increase in physical capacity, improved psychic activity was also been noted, including increased concentration capacity, possibly also increased memory. That is why these drugs are used in geriatric medicine.

SCHISANDRAE CHINENSIS FRUCTUS, Wuweizi The drug is the dry ripe fruit of *Schisandra chinensis* (Turcz) Baill. and *S. sphenanthera*, Rehd. et Wils. belonging to the family *Magnoliaceae*, growing in Northern China, Eastern Siberia and Tibet. A series of dibenzo - (a, c) - cyclooctene derivatives (schizandrines and gomisins) were elucidated as the active principles.

Several studies have shown that this drug increases resistance against the common cold. The drug facilitates the absorption of phosphate and gives a more effective conversion of energy. It stimulates the blood circulation and functions of the brain. Cells receive more oxygen and more easily adjust, as an example, to high altitude. It also has a protective effect on the liver by stimulating the formation of enzymes that destroy the toxins for the benefit of hepatitis patients.

Furthermore, the drug has a normalising effect on blood pressure and can improve sight in terms of adaptation to darkness.

GINSENG ROOT, *Ginseng radix* Ginseng root is, like rhubarb and 'ma huang' (Ephedra), one of the very old drugs within Chinese traditional medicine, which still is used, but now with a scientific background that earlier was missing. (Fig. 24)

Ginseng root consists of the dried roots of *Panax ginseng* C.A. Meyer, family *Araliaceae*. Ginseng root is produced by cultivation, principally in Korea, but also in

Fig. 24. *Panax ginseng* C.A. Meyer. Plant with leaves and fruits. Washed, whole root.

China and Far Eastern Russia, the plants taking 4–6 years to reach maturity. Two commercial forms are available: 'white ' Ginseng, the dried root (frequently with the outer skin peeled off) and 'red' Ginseng prepared by steaming the root under pressure before drying.

Red Ginseng contains all the saponins so far isolated from white Ginseng and some others, formed during the steaming process.

Another species, *Panax quinquefolius* L. with partly different constituents, is cultivated in Northern USA and Eastern Canada and is known as American ginseng.

Active constituents in white Ginseng root are mainly the ginsenosides (in DAB and Ph. Helv. VII min. 1.5%), which are saponin glycosides based on two tetracyclic dammarane triterpenes: 20 (S) protopanaxadiol (e.g. ginsenosides Rb_1, Rb_2, Rc, Rd) or 20 - (S) - protopanaxatriol (e.g. ginsenosides Re, Rf, Rg_1, Rg_2, Rh). Ginsenoside Rb_1 is quantitatively predominant in the diol series; while Rg_1, predominates the triol series.

(Panax ginseng)

(20 S) – Protopanaxadiol R_1 R_2 R_3 = H
(20 S) – Protopanaxtriol R_1 R_3 = H
R_2 = OH

Ginsenosides Ra → Rd
R_1 = Glc (1 → 2) Glc
R_2 = H
R_3 = 2 or 3 sugars
(Glc, Ara. Xyl)
Ginsenosides Re → Rg
R_1 = H
R_2 = 1 or 2 sugar (Glc, Rha)
R_3 = No or 1 sugar (Glc)

As can be seen from the formula, Protopanaxadiol-glycosides have the sugar moieties bound to the OH-groups at C-3 and C-20.

Protopanaxatriol-glycosides have the sugar moieties bound to the OH-groups at C-3, C-6 and C-20. On the other hand there are never any sugars bound to the OH-group in the C-12 position. The sugar components are glucose, arabinose, rhamnose and glucuronic acid.

The Ginseng extract has adaptogenic, stimulant and tonic actions, decribed over the past 30 years, and detailed reviews of published work are available. It has been demonstrated in animal studies that Ginseng produces improved learning, memory and physical capability, resistance to infection and enhancement of energy metabolism.

The adaptogenic effect and the improvement of physical and mental performance in humans have repeatedly been verified by various research groups.

Ginsenosides of the diol type tend to have sedative calming properties while those of the triol type have stimulant effects, therefore the balance between diols and triols can be significant. All ginsenosides have antifatigue effects as shown in the mouse-swimming test.

Ginseng extract is used as a mild stimulant and to diminish infections and as a tonic for physical training, for tiredness and feelings of weakness.

Ginseng is one of the most tested of medicinal plants from a toxicological point of view and it is clear from the studies that it is an extremely non toxic drug. There are some interactions with other medicines, e.g. ginseng may potentiate antihypertensives and warfarin and it may also permit a reduction of Insulin dosage in diabetics because of a slight hypoglycaemic effect. One aspect of Ginseng which has received much attention is the so-called ginseng abuse syndrome (GAS) involving symptoms of hypertension, diarrhoea, nervousness, irritability when high doses of ginseng were consumed. This study has been discredited because the type of ginseng was not identified and because of likely adulteration. However, the GAS is still referred to in many texts.

ELEUTHEROCOCCUS, *Eleutherococci radix.* Eleutherococcus consists of the dried roots and rhizome of *Eleutherococcus senticosus* Maxim. (*Acanthopanax senticosus* (Rupr. et Maxim) Harms), family *Araliaceae.*

The plant *Eleutherococcus senticosus* is a thorny bush that grows in Siberia and Northern China. The English name of the drug has at least three synonyms, 'Spring eleutherococcus', 'Thorny ginseng' and 'Siberian ginseng'.

The active constituents are: Eleutherosides A-G. This group is chemically heterogenous: Eleutheroside A is the ubiquitous phytosterol daucosterol. Eleutheroside B (=syringin) is a phenylpropanoid, whereas Eleutheroside B1 is a coumarin-derivative. Eleutheroside C is methyl-α-D-galactoside; Eleutheroside D and its diastereoisomer Eleutheroside E are lignan-dervatives. Some other eleutherosides are incompletely characterised. The drug also contains polysaccharides, of which the main component is a heteroxylan of MW ca. 30,000.

The actions of the drug are adaptogenic, immunomodulatory, centrally stimulating effects. Substantial pharmacological and clinical research on *Eleutherococccus* has also been carried out in the former Soviet Union (above all by Professor I.I. Brehkman in Vladivostok). These have been verified in Europe and the USA. No side-effects have been observed at normal dose levels.

ROSEROOT, *Sedi rosei rhizoma* The drug consists of the rhizome of *Sedum roseum* (L.). Scop. (syn. *Rhodiola rosea* L.), family *Crassulaceae.*

The plant grows in the European mountain regions, including the mountains in Scandinavia. It contains the glycoside Salidroside, which has a centrally stimulating effect. It has the ability to bind ATP, which is useful in counteracting stress and mental exhaustion.

POLLEN In the preparation of pollen extracts generally, different kinds of pollen are used in order to get as many different bioactive pollen components as possible.

The various kinds of pollen are mixed in definite proportions and the mixture undergoes an enzymatic degradation of the membrane that covers the pores and of the larger allergen molecules to smaller molecules, which are in practical terms non allergic. After this, an extraction of the bioactive pollen components is carried out. In this way the largest possible content of bioactive pollen components and a minimum of allergens is obtained.

Pollen extract contains serveral kinds of organic substances: carbohydrates, sporopollenines, aminoacids, proteins, nucleic acids, alkaloids, carotenes, flavonoids, organic acids, phytosterols and plant hormones.

Furthermore, there are several trace elements such as manganese, cobalt, nickel, copper, zinc and silica.

The field of use and indications for pollen extracts are:

1. To increase physical performance in young and grown-up persons. This is especially for sportsmen and recreation patients.
2. To increase mental capacity in elderly people.
3. Has a mild stimulating effect without a 'kick'.
4. Increases non-specific resistance against external factors.
5. The supporting effect is especially evident in infections and inflammations.

N06B F Psychostimulants that are hallucinogenic Plants that provoke hallucinations have been used by many tribes in all continents for ritual purposes under very different conditions. From a phytochemical point of view this group is very heterogenous. Non-nitrogenous hallucinogens are found in Cannabis and Nutmeg, whereas nitrogenous hallucinogens of an alkaloidal nature are found in the Peyote-cactus, the mushroom *Psilocybe mexicana* with the Mexican name Teonanacatl, the mushroom fly agaric, *Amanita muscaria*, the leaves of the bush Khat, the coca bush and the seeds of climbers, named Ololiuqui. These drugs are dealt with in Chapter VIII.

P. Antiparasitic products

PO1 Antiprotozoals

PO1A Agents against amoebiasis and other protozoal diseases

IPECACUANHA, *Ipecacuanhae radix* The drug consists of the dried underground organs of *Cephaelis ipecacuanha* (Brotero) A. Richard or of *Cephaelis acuminata* Karsten, family *Rubiaceae*, or a mixture of both species.

Fig. 25. *Cephaelis ipecacuanha* (Brotero) A. Richard. Cultivation in Southern Nicaragua.

The drug from *C. ipecacuanha* is known as Matto Grosso Ipecacuanha; from *C. acuminata* as Costa Rica (or Cartagena or Nicaragua) Ipecacuanha. (Fig. 25)

It is a perennial plant, which grows spontaneously in the undervegetation of the rainforests in the Matto Grosso and Minas Geraes provinces in Brazil. Nowadays it is also cultivated in the south of Nicaragua and also in the province of West Bengal (Mungpoo and Rongo).

The active ingredients are the bisbenzylisoquinoline alkaloids emetine and cephaeline which occur in the whole plant; highest concentrations are in the roots with less in the stem and least in the leaves. The ratio emetine: cephaeline is generally 2:1 in Matto Grosso Ipecac and 1:1 in Costa Rica Ipecac.

The ipecacuanha root has three actions: expectorant, emetic and amoebicidal.

The expectorant action is due to the irritant effect of emetine and cephaeline on the gastrointestinal tract, which causes a reflex increase in the secretion of respiratory tract fluid. The dose is 0.5–2 mg of total alkaloids (equal to approx 25 mg of dry root).

As an emetic, 20–40 mg of emetine for adults, the effect is shown after 20– 40 mins. due to local irritation of the gastric mucosa, which produces reflex vomiting and also at a centrally mediated action on the chemoreceptor trigger zone.

The amoebicidal action is used in the treatment of amoebic dysentery, which occurs endemically in tropical and subtropical regions and is caused by a protozoa, *Entamoeba histolytica*. A stronger amoebicidal action is given by the semi-synthetic 2-dehydroemetine. Since emetine is excreted slowly, there is a risk of accumulation.

PO1B Antimalarials

MALARIA is a disease mostly in tropical areas, where it is a major medical problem. Malaria is caused by a parasitic protozoa of the genus *Plasmodium* and is transferred when an infected female mosquito of the genus *Anopheles* bites a person and *Plasmodium* sporozoites enter the blood, where they first reach the liver and develop into merozoites over a period of 5–7 days without giving any symptoms. Then the immature merozoites penetrate the red blood corpuscles, where they divide asexually to form merozoites. When this process is complete, the blood corpuscles rupture and the merozoites enter the blood plasma. The rupture of the erythrocyte membrane provokes a fever, which occurs every second day after infection with *Plasmodium vivax*, every third after infection with *Plasmodium malariae*; after infection with the severe *Plasmodium falciparum* fever is more irregular, because the parasites of this species do not develop simultaneously.

The merozoites, which entered the circulating blood after the rupture of the erythrocyte membrane, partly infect new red blood cells, partly develop to a sexual form: gametes. These gametes can pass to a healthy mosquito when it bites a person suffering from malaria.

The gametes copulate in the gastrointestinal canal of the mosquito. The fertilised egg goes from the intestinal canal to the salivary glands, where, via intermediary stages, they ripen to sporozoids and the circle is completed.

Quinine acts mainly on the merozoites (also called schizonts), and less on the gametes by inhibition of the syntheses of nucleic acids.

Certain forms of malaria, especially infection with *Plasmodium falciparum*, have shown resistance to synthetic drugs but not to quinine; as a result over 2 million people die from malaria and several hundred million are infected each year. For many, the only effective treatment involves either Quinine or Artemisinin.

PO1B C Quinine alkaloids

CINCHONA BARK, *Cinchonae cortex*. Cinchona bark consists of the dried bark of *Cinchona pubescens* Vahl, (*Cinchona succirubra Pavon* Ex Klotzsch), family *Rubiaceae*, or of its varieties or hybrids. For extraction of Cinchona alkaloids other species are extensively used. Four species, and certain hybrids, are of commercial value for cultivation: *C.calisaya* Wedd., *C.ledgeriana* Moens ex. Trimen, and hybrids. The species with high quinine content is *C. ledgeriana*. The cinchona trees grow wild only in the Andes, from Venezuela to Argentina, at a level of 1,000–3,500 m above sea level. Since the beginning of the 20th century cultivated

Fig. 26. *Cinchona pubescens* Vahl (= *C. succirubras Pav.* ex Klotzsch). Cultivation in Province of West Bengal, India.

Fig. 27. *Cinchona pubescens* Vahl (Succirubra Pavan Ex C. Klotsch) Debarking, Rongo, Province of West Bengal, India.

plants have been found in Guatemala, Guinea, Zaire, India (the province of West Bengal) and Indonesia (Java). In cultivation the trees grow for 10–12 years, when they are uprooted, and the bark from stem and root is dried and extracted. (Figs 26, 27)

The European Pharmacopoeia requires a minimum of 6.5% total alkaloids of which 30-60% are of the quinine type (comprising more than 30 compounds). The alkaloids are mainly bound to quinic acid. The most important are two pairs of stereoisomeric quinoline alkaloids, quinine and quinidine, and their demethoxy derivatives cinchonidine and cinchonine. The content of quinine varies in the different species. It is highest in *C. ledgeriana*, where it constitutes approx. 30% of the total content of alkaloids. In the species *C. pubescens* the content of quinine is much lower; instead the stereoisomers cinchonine and cinchonidine dominate with approx. 50% of the total content.

The content of quinidine is low in all species, about 10% of the total content. Since quinidine is frequently used as an antiarrythmic agent, quinine is transformed industrially to quinidine by oxidation and reduction, to quinidine.

The bitter principles of Cinchona bark, which include the alkaloids and also quinovin, which is a mixture of bitter triterpene glycosides: 3-quinovoside (appr. 60%) and 3-glucoside of quinovaic acid (appr. 30%), provoke an increased secretion of gastric juice and saliva, thus stimulating the appetite. Tonic water contains quinine as a stomachic tonic.

QUININE is toxic to several bacteria and protozoas and in therapeutic doses acts as a schizontocide and is used in the treatment of malaria. Since the malaria protozoas now show resistance against the synthetic remedies, mainly chloroquine, natural schizontocides have been the subject of renewed interest.

Quinine and quinidine

PO1B D

QUINGHAO, *Artemisae annuae herba* is the dry aerial part of *Artemisia annua* L., family *Asteraceae*, collected in the fall after the flowers are in full bloom. From this ancient Chinese antimalarial remedy two crystalline substances quinhaosu A and B, with a maximal content (0.7-0.8%) in young leaves at the time of flowering have been isolated.

Quinhaosu B was found to be identical with an earlier known sesquiterpene lactone, artemisinin B, with antimalarial effect and similar mechanism of action as quinine. Clinical tests, performed in China, showed good activity against various forms of malaria, including the dangerous cerebral malaria, caused by *Plasmodium falciparum*, and also against strains of *Plasmodium*, resistant to chloroquine; Arteannuin A and the essential oil in quinghao have no antimalaria effect. Electron microscopic investi-

Fig. 28. *Artemisia annua* L. Cultivation in Vietnam

gations have shown that various membranes in the blood schizonts are modified by quinghaosu B. Because artemisinin is lipophilic it must be formulated as an oily preparation. Various derivatives have been produced which are more water soluble, including an ether called artemether and an ester, sodium artesunate. Quinghao is now extensively cultivated in many areas where malaria is endemic.

PO2 Anthelmintics

Infection with intestinal worms occurs frequently in tropical and subtropical regions. The worms that are pathogenic to humans belong to three classes:

1. Trematodes, among others Schistosoma, which occur in Africa, Latin America and Far East Asia.
2. Nematodes: ancylostomes, ascarids, pinworms, trichines, which are found globally, whereas Filaria is found in tropical and subtropical regions.
3. Cestodes, tape worms, that are found globally.

PO2 B Antitrematodals

This group comprises drugs mainly used for trematode infections such as e.g. schistosomiasis. Various plant extracts have been tested in Africa and Latin America with promising results. e.g. *Phytolacca dodecandra.* L'Hér (Endod or soapberry) which contains saponins which kill the snails which act as vectors for the parasite.

PO2 C Antinematodal agents

In this group agents against ascarids and oxyuris are found.

WORMWOOD *Cinae flores, cinae Semen* Wormwood consists of the dried unexpanded flower-heads of *Artemisia cina*, O.C. Berg et C.F. Schmidt, family *Asteraceae.* It is harvested mainly in Kazakshtan where santonin is extracted in a factory in Chimkent. The drug contains an essential oil with 1,8-Cineole (approx. 80%) but the active anthelmintic substance is L-α-santonin.

The drug and pure santonin are used in paediatric medicine for treatment of ascariasis.

QUASSIA, *Quassiae lignum* is dealt with in the section AO9 B (Amara). It is used as an anthelmintic in cluster for treatment of oxyuris.

Agents against ancylostomiasis

CHENOPODIUM OIL *Chenopodii aetheroleum* This essential oil is obtained by steam distillation of the aerial parts of *Chenopodium ambrosoides* L., var. *anthelminticum* A. Gray; family

Chenopodiaceae. The main constituent (65–70%) is ascaridol, a monoterpene with peroxide-bound oxygen, which is chemically stable.

Ascaridol is an active anthelmintic against ancylostomiasis and ascariasis.

PO2 D *Anticestodals*

MALE FERN (Buckler Fern) *Filicis rhizoma* The drug consists of the dried rhizomes with remaining petioles from *Dryopteris filix mas* (L.), Schott, family *Polypodiaceae*, a fern which grows in Europe, Northern Asia and America.

The active ingredients are phloroglucinol derivatives with two ring systems (flavaspidic acid, desaspidine) and with three ring systems (filix acid).

An ether extract of *filicis rhizoma* or pure desaspidin is used therapeutically as capsules, which have a paralysing effect on the tapeworm, followed by an efficient purgative, so that the paralysed tapeworm and its head are eliminated in faeces. Infection with tapeworm *Botriocephalus latus* = *Diphyllobotrium latum* is obtained by eating certain raw fish (whitefish, schelly, powan, vendace). These species of fish contain one of the developing stages of tapeworm, whereas *Taenia saginata*, and *Taenia solium* are tape worms originating from raw beef and pork respectively. Phloroglucinol derivatives with similar structures as the *Filix* substances are found in the drugs:

KAMALA which consists of the trichomes and glands separated from the fruits of *Mallotus philippinensis* (Lam) Müll. Arg., family *Euphorbiaceae*, a tree in India, Pakistan and the East Indies. Similar phlorogucine-type substances occur in *Koso flos.*, which are the flowers of the Ethiopian tree *Hagenia abyssinica*, J.F. Gmelin, family *Rosaceae*. These two drugs have local use as anthelmintics against tapeworm.

PUMPKIN SEED *Cucurbita peponis semen* consists of the dried, ripe seeds of *Cucurbita pepo* L., family *Cucurbitaceae* or its cultivars. The drug can also be the seeds without husk, *Cucurbitae peponis semen decorticatum*. The anthelmintic substance is the amino acid cucurbitin in a dose of 100–250 mg, followed after 3–4 hours by castor oil as an agent against tapeworm.

PO3 *Ectoparasiticides, incl. scabicides*

CEVADILLA SEED *Sadillae semen* is the seed of *Schoenocaulon officinale* (Schlecht et Cham) A. Gray, family *Melanthiaceae*, which is a Central American bulbous plant, containing about 1% of a mixture of alkaloids called veratrine; the most important one is cevadine. Extracts of the drug are insecticidal and have been used against head-lice, although it is a strong poison as well.

DERRIS *Derridis radix* The drug consists the dried roots of *Derris elliptica* (Wallich) Benth. and *Derris malaccensis* Prain, family *Fabaceae*; these *Derris* species grow in South East Asia, where they also are cultivated. They contain as active ingredient an isoflavonoid rotenone, which has a pronounced insecticidal effect. Rotenone and similar substances have the great advantage that they do not leave any toxic residues.

Fig. 29. *Chrysanthemum cinerariefolium.* Cultivation in Kenya.

Rotenone is also found in other genera of the *Fabaceae* such as *Lonchocarpus* species (cubè root, timbo root) in South America and also in *Thephrosia*-species, used as fish poisons in Equatorial Africa.

PYRETHRUM FLOWERS *Pyrethri flos*
The drug consists of the dried flowerheads of *Chrysanthemum cinerariaefolium* (Trev.) Vis., family *Asteraceae.*

This plant is a perennial herb, about one metre in height, endemic to the Balkans (ex.-Yugoslavia) and now grown in many tropical countries, above all, Kenya. Every part of the plant contains pyrethrins, but the highest content is in the flowers, and only these are used. The terpene esters, which have contact-insecticidal effects, are easily decomposed on exposure to air and light, and pyrethrin I and II have greater effects than cinerin I and II. The pyrethrins and cinerins are non-toxic to man and other warm-blooded animals, but they act on the nervous system of the insects causing convulsion, paralysis and death within 10–15 minutes. In order to increase the stability of the pyrethrum extract, antioxidants (hydroquinone, tannins) are added. Various synergists such as piperonyl butoxide and sesamin can be included to increase the insecticidal effect. Various synthetic and semisynthetic pyrethroids are also used as domestic and agricultural insecticides. Because they degrade so quickly pyrethroids are seen as being environmentally friendly and are used in shampoos and lotions for head-lice (*Pediculus Humanis Capitis*).

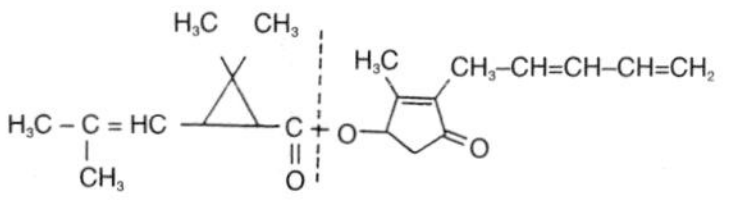

Pyrethrin I

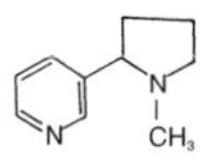

Nicotine

NICOTINE
The waste from tobacco *Nicotiana tabacum* production, family Solanacae, is used for the fabrication of raw nicotine, which also contains nor-nicotine and anabasine. Sulphates are made, and a solution is used as fumigant and insecticide. Nicotine must be handled with great care owing to the high toxicity (for humans); since 40 mg can be a lethal dose.

R Respiratory system

R01 Nasal preparations

R01A Nasal decongestants for topical use

Within classical medicine, nasal drops, nasal sprays or nasal inhalants containing adrenaline-derivatives are used for their vasocontracting effect on the peripheral blood vessels.

A long-established household remedy is the inhalation of menthol vapours. Practically, it is carried out in the following way: Some menthol crystals are poured into a bowl containing very hot water. The menthol crystals are melted and the treated person inhales for some minutes a mixture of menthol and water vapour. A cloth over the head increases the theraputic effect.

Menthol, which is the main constituent of *Menthae piperitae aetheroleum*, acts locally on the skin by stimulation of cold-sensitive nerves, producing a sensation of cold and hyperaesthesia. On the mucosa it produces anaemia and anaesthesia and inhibition of secretion. Furthermore, menthol shows antiseptic and in higher concentration even bactericidal properties. (cf. A01 p46.)

By inhalation a local anaesthetic effect and vasoconstriction of the vessels in the nasal mucosa are obtained.

RO3 Anti-asthmatics

RO3B Other anti-asthmatics, inhalants

ATROPINE Atropine and other tropane alkaloids are dealt with in paragraph AO3B. The use of asthma cigarettes, containing *Stramonii folium*, may be mentioned here.

STRAMONIUM LEAF *Stramonii folium* During smoking 25% of the alkaloids (atropine and related compounds) go into the smoke; 20–80% of which reach the lungs. With an optimal inhalation technique (slowly and deeply) one cigarette gives 0.25 mg atropine in the airways, which gives a strong bronchodilation. Since stramonium and its alkaloids are highly toxic, care must be exercised when using these products.

EPHEDRA *Ephedrae herba* Ephedra under the name of 'Ma huang', ginseng and rhubarb are the three drugs, with the longest history of use in Chinese medicine, dating back about 5,000 years.

Ephedrae herba consists of the dried aerial parts of xerophytic *Ephedra* species belonging to the family *Ephedraceae*. The plant is a shrub, 0.5–1 m high with scale-like leaves that clasp the stem.

Nowadays the *Ephedra* species are mainly used for extraction of the active ingredient, the alkaloid (–) ephedrine. Not only the classical Chinese species *E. sinica* Stapf. is used, but mainly the species *E. gerardiana*. Wall ex. Stapf and *E. major* Host. (= *E. nebrodensis* Finco), which both grow in India and Pakistan, and *E. distachya* L., which grows in Southern France.

The alkaloid ephedrine was isolated in 1885 by the Japanese chemist W.N. Nagai. Two pharmacologists, the Chinese K.K. Chen and the American G.F. Schmidt, showed in the 1920s that ephedrine is a sympathomimetic, i.e. it has an adrenaline-like effect with two important modifications: it is much less active, but the effect has a longer duration and it is effective by oral administration. Owing to the bronchodilating action, ephedrine is used in the prophylactic treatment of asthma. The vasoconstrictor effect explains its use in allergic nasal catarrh and for pretreatment prior to spinal anaesthesia. Together with antitussives (see RO5) ephedrine is administered for asthmatic cough. Furthermore, ephedrine exerts a CNS-stimulating action (see Chapter VII).

Since ephedrine has two asymmetric C-atoms, there is, besides L-ephedrine (which has the strongest pharmacologic effect), also (+)pseudoephedrine, which has an α-adrenergic action, whereas L-ephedrine stimulates both α- and β-receptors. Ephedrine acts directly on the receptors, but mainly by liberation of endogenous noradrenaline, which in turn acts on the receptors.

The enantiomeric forms D-ephedrine and (–) pseudoephedrine do not occur in nature. Concern has been expressed about the indiscriminate use of Ephedra for its stimulant and anorectic effects (Chapter VIII).

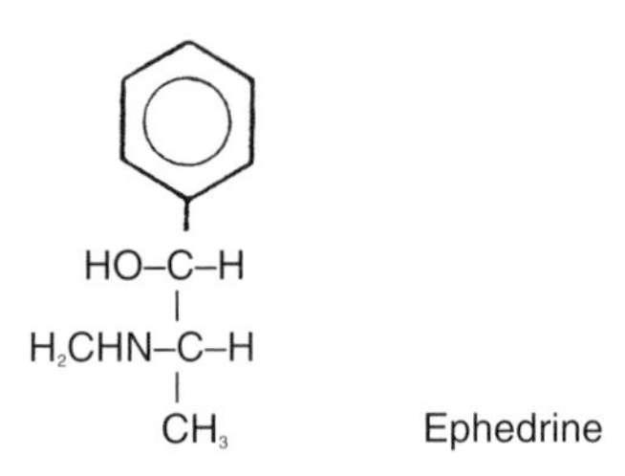

Ephedrine

RO3D Other anti-asthmatics for systemic use (intervaltherapy)

THEOPHYLLINE Theophylline, caffeine and theobromine are the three purine derivatives that are used therapeutically. As a group they are dealt with under NO6B. Among the three alkaloids, theophylline has the strongest brochodilating action.

Theophylline gives with ethylene diamine (2:1) a water-soluble salt, aminophylline, which is given by slow intravenous (not intramuscular) injection in a dose of 250–500 mg. Oral administration for a lengthy period may cause irritation of the gastric mucosa with nausea, vomiting and pains in the upper abdomen.

KHELLIN Khellin is extracted from the schizocarp (*Ammeos visnagae fructus*) of *Ammi visnaga*, L. (Lam.), family *Apiaceae* (see C01D), and has a vasodilating effect. It is an ingredient in asthma preparations. Some side-effects have limited its clinical use. It served as a model compound for the synthesis of sodium cromoglycate.

RO5 Cough and cold preparations

RO5C A Expectorants According to an accepted pharmacological definition expectorants are drugs that influence the properties, formation and transport of the bronchial secretion.

The plant expectorants have long been used empirically. The mode of action is based on three mechanisms:

1. A decrease in viscosity of the secretion; where herbal teas are used this also occurs because of the addition of hot water.
2. A gastropulmonary reflex: Irritation of the upper gastrointestinal tract produces emesis: afferent impulses go to the emetic centre in visceral sensible nerves in the gastric mucosa. Emesis starts with saliva secretion and indisposition. The dose of the 'reflex expectorants' is so chosen so that the first stage is reached: secretion of less viscous fluid in the bronchial cells.
3. A direct action of essential oil on the bronchial glands: the serous glands are selectively stimulated and the mucous glands are inhibited, the result being a (almost non-viscous) secretion.

The decrease of viscosity is obtained by splitting the disulphide bridges in the proteins of the mucilage. Furthermore, surface active saponins decrease the viscosity of the mucilage by decreasing the surface tension.

a) Aromatic expectorants They are administered orally, rubbed into the skin or inhaled. After resorption they are excreted partly through the lungs and stimulate (direct inhalation) the bronchial secretion so that the secretion is easily liquified. Furthermore, they have a spasmolytic effect on the bronchioles (the smallest ramifications of the bronchi).

THYME *Thymi herba* The drug consists of whole leaves and flowers separated from the previously dried stems of *Thymus vulgaris* L. or *Thymus zygis* L. or a mixture of both species, belonging to the family *Lamiaceae (Labiatae)*.

Thyme is commonly used as a culinary herb and is characterised by its volatile oil (0.8–2.6%). Phenols are the main ingredients, thymol in *Thymus vulgaris* and carvacrol in *Thymus zygis*.

The lowest content of volatile oil (according to DAB 8) is 1.2% of which at least 0.5% should be phenols.

Thymol has an antiseptic effect (25 times stronger than phenol), antispasmodic, antitussive, expectorant, and astringent properties, and the flavonoids also have activity as antispasmodics in bronchial muscle.

ANISEED *Anisi fructus* The drug consists of the schizocarpic fruit of *Pimpinellla anisum* L., family *Apiaceae*, a herb endemic to the Eastern Mediterrean Region, and cultivated in Southern Europe. Aniseed contains 1–3% volatile oil, *Anisi aetheroleum*. The main ingredient is trans-anethole (80–90%). *Anisi aetheroleum* can also be obtained by steam distillation of

STAR-ANISE *Anisi stellati fructus* The fruit of *Illicium verum*. Hook f., family *Magnoliaceae*, a tree in South East Asia.

The fruit of star-anise contains 2.5–5% volatile oil.

Both aniseed and star-anise are used as spices and for distillation of anise oil. Besides the main ingredient anethole, it contains a polymeric product, dianethole, with oestrogenic effects. Aniseed has been reputed to increase milk secretion, promote menstruation, facilitate childbirth and increase libido. Aniseed is reported to be expectorant, carminative and exert sympathomimetic-type effects. These effects are attributed to anethole. The expectorant effect of aniseed oil is experimentally verified by measuring the amount of secretion from an upper respiratory tract and measuring its viscosity. In therapeutic doses, anethole is reported to cause minimal hepatotoxicity.

FENNEL FRUIT *Foeniculi fructus* The schizocarpic fruit of *Foeniculum vulgare*, Mill., family *Apiaceae*, which is a herbaceous plant, endemic to the Mediterranean aera. The fennel fruit contains 2–6% volatile oil, *Foeniculi aetheroleum*, which has a bitter camphor-like taste of the optically active (+) fenchone (called BITTER FENNEL, *Foeniculi amari fructus*,) with minimum 60% anethole and 15%, fenchone, and the SWEET FENNEL, *Foeniculi dulcis fructus*, with trans-anethole 80%.

Like Aniseed, Fennel has a secretolytic, spasmolytic, carminative and antiseptic effect with the same use as anise.

b) Expectorants containing saponins Saponins are a group of glycosides which lower the surface tension of water solution (foam producing) and have a haemolytic effect (destroy the erythrocyte membrane so the haemoglobin goes into the surrounding water solution).

The saponins have either a steroidal-aglycone or a triterpenoidal aglycone. Only the last-mentioned group of saponins are used as expectorants.

The triterpene aglycones have a C-30-skeleton, which have five rings (pentacyclic), more rarely four rings (tetracyclic). Most triterpenes, are derived from an oleanane hydrocarbon, and only a few have an ursane or dammarane skeleton. The most common triterpenoids are β-amyrin, oleanolic acid and ursolic acid.

The acidic nature of the triterpene saponins is due to a carboxyl group in the aglycones or in the sugar-part, if there is an uronic acid.

Those saponins used as expectorants have low toxicity, because they are not absorbed intact. Haemolysis cannot be the primary cause of death, when rats, after a lethal dose of primulic acid A (active ingredient in *Primulae radix*), live 5–7 days after administration.

QUILLAIA BARK *Quillajae cortex* is the bark of *Quillaja saponaria*, Molina, family *Rosaceae*, from which the cork is removed. This species is an evergreen tree in Chile (in the valleys of the Cordilleras), Peru and Bolivia. The drug consists of the inner bark which is generally cut before marketing. It has an acrid taste and causes sneezing. The bark contains about 10% of a saponin mixture (Quillaia saponin), which besides its use as an expectorant also has a technical use and as an adjuvant in some vaccines as the saponin potentiates the immunising power of the vaccine.

COWSLIP (ROOT) *Primulae radix* is the root of *Primula veris*, L. and *Primula elatior* (L.) Hill, family *Primulaceae*, which grows in Asia and Europe (South Russia to England), collected in the autumn.

The drug contains 5–10% saponin (higher content in *Primula veris* than in *Primula elatior*) with the main saponin Primulic acid A, its aglycone being Primulic genin A, and the sugars are arabinose, galactose, glucose, glucuronic acid, rhamnose and xylose.

In vitro the saponins have documented ability to inhibit prostaglandin (PG)-synthetase, but to a lesser extent than aspirin.

Primulae radix is a typical reflex-expectorant: the effect is a weak emetic effect, i.e. stimulation of sensitive nerves in the stomach mucosa stimulates by reflex the branches of *vagus* nerve that supply the bronchi. Antibacterial and antiinflammatory effects are also found.

SENEGA *Senegae radix* (*Polygalae radix*) is the dried root and rhizome of *Polygala senega* var. *senega* and/or var. *latifolia* Torrey et A. Gray, family *Polygalaceae.* These plants grow in Southern Canada and the Middle United States, in the Mississippi region. Nowadays it is also cultivated in Japan.

Senega root has a typical smell due to its methyl-salicylate content. It contains 6–10% triterpenoid saponins, related to senegin II, the major component of the mixture present in the var. *latifolia.* They all have a glucose attached to C-3, but differ in the oligosaccharide chain that is ester bound at C-28.

Senega has expectorant, diaphoretic, sialogogue and emetic properties. It is used specifically for chronic bronchitis.

LIQUORICE *Liquorice root* (*Liquiritae radix*) and liquorice extract (*Glycyrrhizae extractum crudum, Succus liquiritiae*) are obtained from *Glycyrrhiza glabra* L., family *Fabaceae.* Liquorice root comprises root and underground stems (stolons). Three varieties are known: *var. typica*, a perennial herb, 1–2 metres high, growing in southern and central Europe; *var. glandulifera*, growing in central and southern Russia; *var. violacea*, growing in Iran, Iraq and Afghanistan.

Glycyrrhiza glabra is cultivated in Spain, Italy, France, Germany and the United States. The Russian liquorice root is peeled, the Spanish unpeeled. The crude drug from Russia and the Middle East comes from wild-growing plants; the roots and stolons are harvested from plants, generally 3–4 years old.

Liquorice extract is mainly produced in southern Italy where the crude drug is extracted with water, and the extract is filtered and evaporated under vacuum to a viscous mass, which is cast into blocks or sticks and gradually solidifies. The colour is brownish black and it has a shiny fracture and a sweet taste.

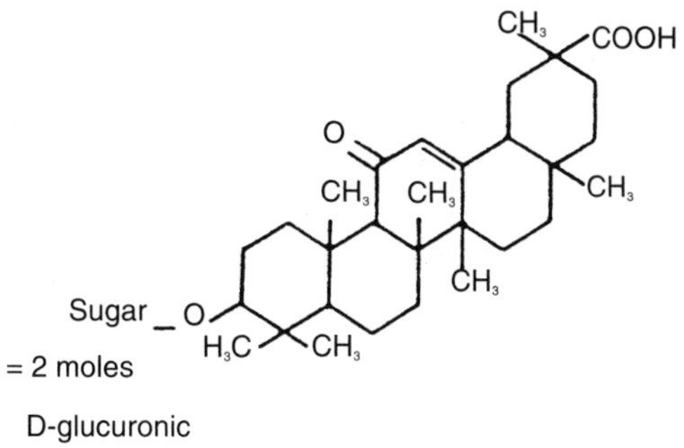

Sugar = 2 moles D-glucuronic acid

Glycyrrhizin

The saponin is glycyrrhizin and consists of the aglycone glycyrrhetinic acid and two molecules of glucuronic acid. It is about 50 times sweeter than sucrose, but the aglycone has no sweet taste. The content of glycyrrhizin is 8–12% in the drug and 20–25% in the extract.

The liquorice extract is a classical expectorant component in cough syrups. Furthermore, liquorice is used, owing to its sweet taste, in the confectionery industry and also the food industry (liquorice ice cream!). After overdosage of liquorice (more than 25 g), cortisone-effects (due to the structural similarities between cortisone and glycyrrhetinic acid, among others the 11-keto-group) are obtained: oedema, salt retention, hypokalaemia (low potassium content in blood).

The liquorice root contains furthermore the flavonoids liquiritoside and isoliquiritoside, the aglycones of which have a spasmolytic effect (about half as active as papaverine), which is used for treatment of ulcer. The preparations (see AO2BX) containing deglycyrrhizinised liquorice have a reduced effect mineralocorticoid. The exact mode of action is yet unknown, but increased production of mucin, increased production of cells in the gastric mucosa and a spasmolytic effect have been demonstrated.

c) Alkaloid-containing expectorants

IPECACUANHA *Ipecacuanhae radix.* This drug is described in section PO1 and is used in low doses as an expectorant. The alkaloids emetine and cephaeline are irritant to the gastrointestinal tract, causing a reflex increase in the secretion of respiratory tract fluid.

COCILLANA (Syn. Guarea, Guapi Bark) *Guareae Cortex* is the bark of *Guarea rusbyi* (Britt.), Rusby and closely related species, family *Meliaceae.* It is obtained mainly from Bolivia. Cocillana is a more stimulating expectorant than Ipecacuanha. The unidentified alkaloid fraction contributes most probably to the expectorant effect.

RO5C B Mucolytics

ADHATODA The leaves of *Adhatoda vasica*, Nees, family *Acanthaceae*, are used in India. The constituents include the alkaloid vasicine (peganine), which has a mucolytic effect, i.e. lowers the viscosity of the bronchial mucus by depolymerisation and makes it more liquid, so that it can be transported. The mucocilliary transport of mucous to the periphery depends on the mucous having certain rheological properties. The drug is used as mucolytic in cough mixtures in the form of a liquid extract.

Vasicine has been a model substance for the synthesis of bromhexine.

RO5 D Antitussives

A cough is one of the most common symptoms of illness in the respiratory tract. Very often it is very disturbing (troublesome) for the patient and is an important reason for the extensive use of cough medicines.

A cough is characterised as a forced, audible, blow of air through the mouth; thus, a highly increased expiratory velocity of the stream of air. This can be achieved by a corresponding increase in the expiration. Usually the expiration takes place passively as an after-effect of the inspiration activity of the respiration centre. If the expiration is strongly increased, the respiration centre must be expiratory active. Such activity also occurs on other occasions; i.e., after hyperventilation during strong physical activity. According to this hypothesis, a cough is a special form of activity of the respiration centre, along with others.

The antitussives can be divided in two categories:

1. The phytomedicines that inhibit the cough-centre: Codeine, glaucine, noscapine and semisynthetic ethyphine.
2. The phytomedicines that inhibit the cough-reflex: Mucilaginous herbs such as Althea root and Icelandic moss.

1. Phytomedicines that inhibit the respiration centre: The alkaloids codeine and noscapine are found as natural ingredients in the drug opium, obtained from the opium poppy, *Papaver somniferum*, L., family *Papaveraceae.*

The opium poppy contains four alkaloids: papaverine, morphine, codeine and noscapine (=narcotine), with four areas of therapeutic use:

RO5D:	Cough inhibition:	codeine, noscapine
AO3A:	Spasmolytic:	papaverine
AO7:	Anti-diarrhoea effect:	opium
NO2A:	Analgesics:	morphine, codeine

The production of opium and the cultivation of the opium poppy is described in section NO2A.

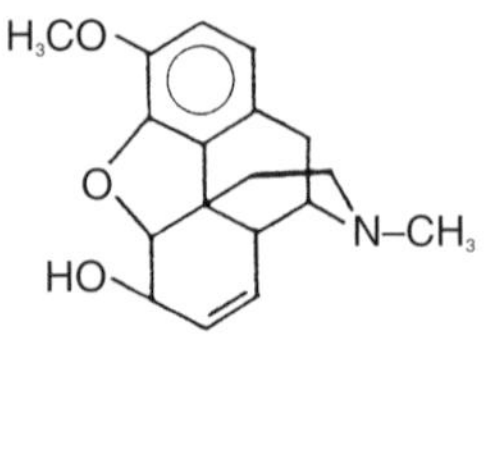

Codeine

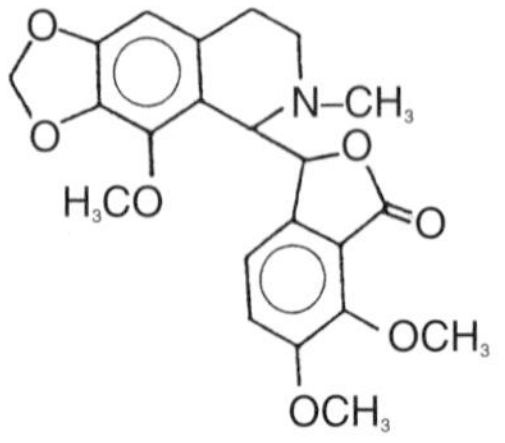

Noscapine

CODEINE Codeine, which is a methly-ether of morphine, is found in opium at a level of 0.3-3%, but is produced semi-synthetically by methylation of morphine. It is used as water-soluble codeine phosphate (which contains 70% codeine-base).

The usual dosage is 30 mg × 3-5 with maximal effects within 1-2 hours and duration of 4-6 hours.

The metabolism and elimination are slow and the metabolites are: unchanged codeine (about 10%); nor-codeine (main metabolite); morphine (about 10%), mainly as conjugates: it can be found 48-72 hours after administration.

Children are generally more sensitive than adults (codeine should not be given to children less than $\frac{1}{2}$ year old). The main risk of poisoning is the CNS-effect and respiratory depression, which can come late (especially for ethyl-morphine). An effective antidote is Naloxone (Narcan)®

NOSCAPINE (NARCOTINE) As an optically active, laevorotatory substance noscapine is found in opium (2-12%). This alkaloid is, next to morphine, the most plentiful of the opium alkaloids. Structurally it differs from the phenanthrene derivatives morphine and codeine by its phthalide isoquinoline-skeleton.

Noscapine is used as the water-soluble hydrochloride, which has a bitter taste. Noscapine lacks the analgesic, respiratory depressive and constipating effects of codeine. Its antitussive action is weaker than that of codeine. The absorption from the gastrointestinal tract is good, the duration is about 4 hours. Normal dosage is 25-50 mg three-four times daily and maximal daily dose is 250 mg. After overdosage the symptoms are headache, dizziness, lethargy, and vomiting. Toxic doses may give negative inotropic effects on the heart.

GLAUCINE Glaucine is the main alkaloid in *Glaucium flavum*, L., family *Papaveraceae*. Glaucine is extracted from this plant, cultivated in Bulgaria. It has an antitussive effect, but has no contractive action on the bile duct. Lately glaucine has had an increased use.

ETHYL MORPHINE This semisynthetic derivate has an ethyl group instead of the methyl in codeine and is used on a large scale in cough syrups.

2. Phytomedicines that inhibit the cough-reflex Alleviation of a cough can be achieved with mucilages, especially if the irritation of pharyngeal mucosa is the cause of unproductive cough. In cases of laryngitis (inflammation of larynx) and pharyngitis (inflammation of pharynx) the mucosa cannot fulfil its normal function of protection: a constant production of mucosal secretion. The intake of 'cough-tea' containing mucilages or sucking on cough-tablets diminishes the irritability of the mucosa and thus of the cough attack.

MUCILAGES Mucilages are substances based on less well-defined polysaccharide derivatives, which locally exert an inhibition of irritation by covering the mucosa with a protective layer.

Not all mucilage-containing drugs are used for this purpose. In traditional medicine two drugs are used:

MARSHMALLOW ROOT The root consists of the dried, peeled root of *Althaea officinalis*, L., family *Malvaceae*, collected in the autumn from plants at least two years old.

It is a perennial 1-2 m tall herb with an erect woody stem, which grows in Central Europe. Late in the autumn the roots from wild and cultivated plants are harvested and the cork and outer parts of the bark are removed. The peeled root is dried quickly at ca 40°. The crude drug is often cut into small cubes, to facilitate the production of starch-free mucilage, which is made by maceration with cold water of the root cubes.

The content of mucilage is about 15% in late autumn and mucilage fractions include an acidic polysaccharide with MW of ca. 30,000 composed of L-rhamnose, D-glucuronic acid; also an arabinan with a highly-branched structure of about 90 α-L-arabinose residues and three predominantly linear glucans composed of 1–6 linked α-D-glucopyranose units.

The pharmacological effects of marshmallow root are due to its content of mucilage, which forms a protective layer on mucous membranes or skin, soothing irritation and inflammation. It is used topically: as a mouthwash or gargle for inflammation of the mouth and pharynx. It is a component in tablets, tea, and mixtures.

ICELAND MOSS Iceland moss consists of the dried thallus of *Cetraria islandica*, L., Acharius, family *Parmeliaceae.* Iceland moss is odourless, with a taste both mucilaginous and distinctly bitter. It is supplied as pieces of lichen branches and collected from the wild in subarctic or mountainous regions of Europe and obtained mainly from Scandinavian and Balkan countries, Russia and also Canada.

It contains more than 50% of a mucilage which is soluble in warm water and consists of two polysaccharides, lichenin, which is built of 60–200 glucose molecules, and isolichenin, which is soluble in cold water and built of about 40 glucose molecules. The bitter taste is due to the occurrence of lichen acids.

The drug is used in bronchitis, lack of appetite and gastric catarrh.

RO 5 X Other cold combination preparations

Commercially available 'common cold' preparations may contain Menthol, Camphor, Turpentine, Clove oil, Thymol, Eucalyptus oil, yellow soft paraffin oil, and that essential oils that evaporate at the temperature of the human body. When the ointment is rubbed on the skin, a local stimulation of the peripheral blood vessels is obtained, which helps to increase the patient's well-being during bouts of the common cold.

S Sensory organs

SO1 Ophthalmologicals

SO1E Antiglaucoma preparations and miotics

For the treatment of glaucoma there are substances available with different modes of action. Two short-acting cholinergics which are used are pilocarpine, which is a first-hand remedy, and physostigmine.

Fig. 30. *Pilocarpus microphyllus* Stapf

PILOCARPINE From the crude drug Jaborandi *Jaborandi folium*, which is the leaflets of *Pilocarpus* species, family *Rutaceae*, the alkaloid pilocarpine is extracted. The Pilocarpus species are shrubs or trees, native to South and Central America and to the West Indies. The most important species at present is *Pilocarpus microphyllus* Stapf, (Fig. 30) which grows in Brazil and gives the drug Maranham-jaborandi containing 0.7–0.8% pilocarpine. Other kinds of jaborandi with larger leaves (but lower pilocarpine content) are Pernambuco-jaborandi from *P. jaborandi* Holmes and Paraguay jaborandi from *P. pennatifolius* Lemais.

Pilocarpine is used in the treatment of glaucoma to decrease the intraocular pressure through increased drainage (Canalis Schlemmi is dilated) over a period of hours.

Maximal effect is obtained by a 4% solution. Using different vehicles the time of contraction can be prolonged and thus the duration of the decrease in the pressure.

Pilocarpine

Physostigmine

PHYSOSTIGMINE (ESERINE) *Physostigmatis semen, Calabaricae semen*, (Physostigmine is extracted) from the calabar bean, . The Calabar bean is the seed of *Physostigma venenosum*, Balf., Fig 31 family *Fabaceae*, which is a climber in Cameroon, Nigeria and adjacent countries. The fruit is a

Fig. 31. *Physpostigma venenosum* Balf. Ripe and unripe pods. Cameron

pod with 1–3 kidney-shaped seeds, which are 2–3 cm long. In the seeds alkaloids are located under the seed shell, the main alkaloid being physostigmine.

The Africans have long used the calabar bean as an ordeal poison. A person who was suspected of being guilty of a crime was forced to eat the crushed beans. If he survived, he was considered innocent. The bark of *Erythrophleum guineense* has also been used for this purpose

The mode of action of physostigmine involves an inhibition of the enzyme acetyl cholinesterase. Physostigmine is an alkaloid with tertiary nitrogen, which gives a good penetration through the cornea.

Traditionally it has been considered that a directly acting remedy like pilocarpine and an indirectly acting remedy like physostigmine have additive effects.

The therapeutic use is (like pilocarpine) to decrease the intraocular pressure in the eye in glaucoma. Furthermore, intestinal peristalsis is stimulated and physostigmine is used to remove intestinal atonia after operations. The effect of physostigmine on striated muscles is also important. As an acetylcholine inhibitor physostigmine (and synthetic analogues) is used in the treatment of myasthenia gravis and as an antidote in the case of overdosage of tubocurarine used as a muscle relaxant in complete anaesthesia.

CANNABIS There is evidence that Δ9-THC from Cannabis may be useful in glaucoma, but there are difficulties in formulation and rebound effects make the therupeutic use of Δ-9-THC or cannabis unlikely.

SO1F *Mydriatics and cycloplegics*

ATROPINE (SEE AO3) Atropine is a long-acting anticholinergic, which is used therapeutically in cases of iritis (inflammation of iris) and iridocyclitis (inflammation of iris and corpus ciliare), but for diagnostic purposes a short-acting synthetic preparation is used, as for instance tropicamide.

SO1X *Other opthalmologicals*

BILBERRIES *Myrtilli fructus*, are the fruits of *Vaccinium myrtillus* L., family *Ericaceae*. There are preparations containing anthocyanosides (25%) from bilberries. Studies indicate positive benefits in patients with diabetic and senile retinopathies and other degenerative conditions of the eye for which it is claimed that they accelerate the accommodation of the eye to darkness. In Italian investigations a dose of 480 mg/day was used.

Chapter VII
CHEMOPREVENTION USING PHYTOCHEMICALS

Introduction

It is estimated that in the developed world, most deaths are due to two major causes; circulatory diseases (heart attacks and strokes) and cancer. A large proportion of these two diseases is attributable to environmental and lifestyle factors including diet, social status, cultural practices, tobacco smoking and alcohol abuse. In recent years health professionals and governments have begun to closely examine ways in which risk factors for these major killer diseases can be modified so as to reduce excess premature mortality from them.

Cancer

Cancer is a term describing a group of diseases all of which involve a state of uncontrolled, malignant growth of cells which are poorly differentiated. Cancer can occur in any organ of the body and there are believed to be over 100 distinct types of cancer, each of which is essentially distinct with no apparent common features.

Research into the chemotherapy of cancers is based on 3 different aspects:

(1) Tumour initiation, i.e. aspects of cell growth and division.
(2) Angiogenesis, i.e. the mechanism by which a tumour develops a blood supply network.
(3) Metastasis, which is the mechanism by which some of the cancer cells become detached from the parent tumour and invade other parts of the body setting up new tumour sites.

Many thousands of compounds have been tested for anticancer activity in the numerous screens and assays which have been developed by the pharmaceutical industry. Some compounds are synthetic, some produced by bacteria or fungi, some are isolated from animals and many are of plant origin. Of the latter, a number have become clinically useful antitumour agents such as Vincristine, Taxol®, Etopside etc. (see chapter VI: L) but a large variety of plant compounds are known to affect tumour cells in different ways, even though they do not to have the potency to become novel drugs. From this we can deduce that plants through their biosynthetic capacity can produce an array of molecules which can influence tumour initiation, promotion or spread. On the other hand some phytochemicals may even cause cancerous growths.

These anticancer compounds are not confined to exotic rainforest plants but can be found even in everyday foods, plants, herbs and spices. This has led to the suggestion that certain food plants and the secondary metabolites they produce could be used to prevent cancer and also other major killer diseases such as heart disease. The whole concept of chemoprevention has received a lot of attention in recent years.

Chemoprevention has been defined as the use of nutrients or pharmacological agents to enhance intrinsic mechanisms that protect the organism against the development and progression of disease, e.g. malignancies and coronary heart disease.

Examples include the use of:

- Low dose aspirin which has been shown through its antiplatelet activity to reduce strokes and heart attacks
- Cholesterol-lowering drugs such as clofibrate and the statins
- Phytochemicals in the diet to reduce cancer risks, e.g. broccoli contains sulpharaphane which blocks metabolic activation of carcinogens
- Green tea contains antioxidant polyphenols which can reduce the risk of a number of diseases
- Particular attention has been paid to lignans, isoflavonoids and procyanidins because of their important preventive role.

Cardiovascular disease

A number of plants and phytochemicals have attracted attention for their ability to reduce many of the risk factors associated with cardiovascular disease. Research into these diseases has shown the relationship between lesions, fatty streaking and plaque formation in blood vessels and the development of strokes and myocardial infarctions. These effects are linked to levels of plasma lipids which comprise triglycerides, cholesterol and other fat substances. It is known that the biosynthesis of lipids involves the condensation of several molecules of acetylcoenzyme A and malonylcoenzyme A in a gradual process of elongation of the fatty acid chain involving the sequential addition of two carbon units giving rise to fatty acids such as lauric acid (12 carbons) and eventually to palmitic acid (16 carbons). Palmitic acid is the precursor of longer chain saturated and unsaturated fatty acids formed through the action of enzymes called elongases and desaturases. Plant cells can produce a variety of polyunsaturated fatty acids via desaturases but mammalian cells cannot. Dietary fat thus becomes important because humans need to consume polyunsaturated fats in the diet to provide fatty acids such as linoleic acid (18 carbons and two double bonds i.e. 18:2) and linolenic acid (18:3) which are required for the production of essential phospholipids. Thus these unsaturated lipids play a key role in the elaboration of membranes and eicosanoids. Prostaglandin, thromboxane and leukotriene synthesis is also dependent on the modification of these acids, to arachidonic acid (20:4) which is a key precursor in their synthesis and as such is central to a number of disease processes. The lipids are transported in the blood bound to protein, called lipoprotein. The fat in food is absorbed in the intestine as chylomicrones, which in the liver are transformed to VLDL particles (very low density lipoprotein) after splitting off the triglycerides. HDL (high-density lipoprotein) is formed in the liver and takes cholesterol back to the liver to be degraded and excreted via the bile. The ability to remove cholesterol from plasma in the form of HDL-cholesterol ('good-cholesterol') is a key to reducing one risk factor for heart disease. This is because cholesterol can be deposited as a plaque (atheroma) on blood vessel walls leading to atherosclerosis where the atheroma causes narrowing and ultimately blockage of the arterial wall leading to coronary heart disease, hypertension and myocardial infarction. Thus prevention of cholesterol biosynthesis through the use of inhibitors such as the statins (lovastatin etc.) either isolated from fungi or found in red yeast (*Monascus purpureus*) Went is widely recognised as a chemopreventive measure.

Polyunsaturated fats of vegetable and fish origin have also attracted interest because of their ability to lower plasma cholesterol levels

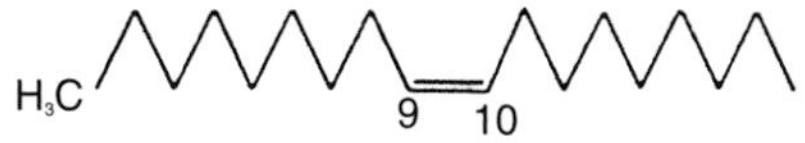

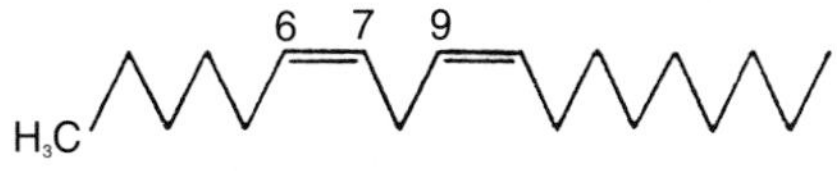

α - Linoleic acid 18:2, w – 6

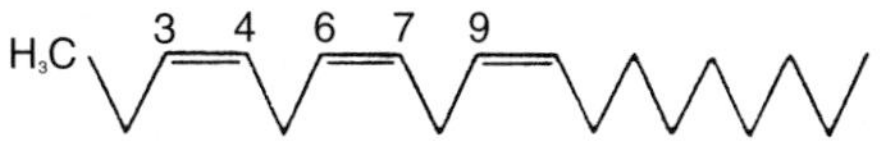

α - Linoleic acid 18:3, w – 3

and to interfere with platelet aggregation which is another important contributor to heart disease.

The existence of atheromatous plaques on blood vessels can initiate a series of changes in blood platelets which leads to the aggregation and the development of thromboses. These changes in platelets are influenced by the balance between pro-aggregation and aggregation-inhibitory forces within the platelet (also known as a thrombocyte) and these in turn are influenced by the metabolism of fatty acids. Thromboxane A2 (TXA2) is synthesised in platelets from arachidonic acid (20:4); it has the strongest platelet aggregating effect among the eicosanoids and is also vasoconstricting. In blood vessels prostacyclin I_2 (PGI_2) is also produced from arachidonic acid and has an opposite effect to TXA_2 in that it inhibits platelet aggregation and acts as a vasodilator. The formation of gamma linolenic acid (GLA) from linoleic acid is important. A decreased formation of GLA and dihomogammalinolenic acid leads to a decreased synthesis of PGE_1 which is the antithrombotic prostaglandin. Increased intake of GLA facilitates the formation of dihomogammalinolenic acid and PGE_1. Intake of essential fatty acids increases the production of prostaglandins which is also stimulated by minerals such as selenium and zinc and by vitamins such as B_3, B_6, C and E.

The cholesterol lowering effect of GLA is considerably larger than that of linoleic acid indicating that linoleic acid must be transformed to GLA to have any effect on cholesterol metabolism. It is suggested that GLA should be combined with fish oils to maximise the beneficial effects on cholesterol and blood lipid metabolism. When saturated fats are replaced by polyunsaturated lipids from fish in the diet of volunteers a decrease in systolic blood pressure can be recorded. The favourable effect of fatty acids such as eicosapentaenoic acid (EPA) and docosahexaenoic acid (DHA) in fish oils is partly antithromobotic because they are less likely to result in TXA2 synthesis and partly anti-arteriosclerotic through a reduction in blood pressure and in plasma lipids in general. It has been established that monounsaturated (e.g. oleic acid 18:1) and polyunsaturated fatty acids are able to lower cholesterol levels (particularly the LDL or low-density lipoprotein form). General dietary advice indicates that saturated animal fat intake should be reduced to less than 10% of total calories each day. Increased consumption of oils such as olive, corn, soybean, sunflower and safflower within an overall total fat consumption of 30% of total calories is generally recommended. Thus dietary modification including increased consumption of fibre-rich foods and antioxidants is increasingly encouraged.

Medically there is an acceptance of the value of pharmacological interventions designed to prevent disease development by using cholesterol/lipid lowering agents and low dose aspirin to prevent platelet aggregation. In terms of phytochemical and herbal materials most attention has focused on the value of Garlic as an hypercholesterolaemic/hyperlipidaemic agent (see section VI: BO4) and on the role of flavonoid-based antioxidants. Of particular importance are dimeric flavonoids known as the biflavonoids, and their related higher molecular weight derivatives called procyanidins.

The procyanidins are noted for their effect on the cardiovascular system either directly

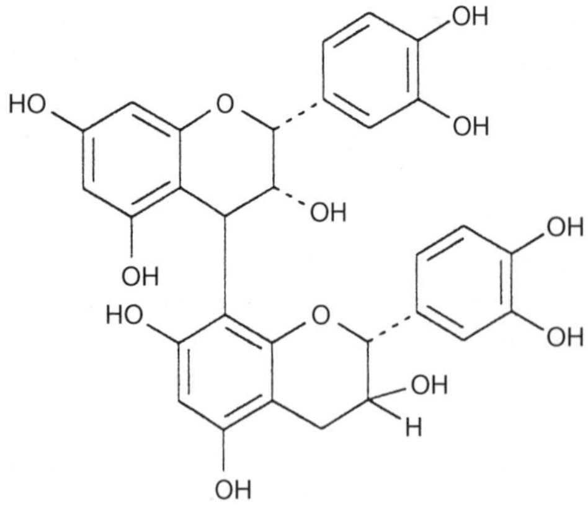

Procyanidin B1

or through an effect on atherosclerosis and cholesterol levels. Procyanidins are responsible for the effects of Hawthorn - *Crataegus monogyna* and are found in the leaves, flowers and fruits. The effectiveness of Hawthorn extracts and compounds isolated from them has been confirmed in animal studies and subsequently in double-blind clinical trials. (cf. CO1Ep76.).

The French paradox

These procyanidins occur in other plants and attention was first drawn to this in 1979 when a study of mortality from Coronary Heart Disease (CHD) in a number of Western countries was performed. It was noted that populations having the lowest mortality due to heart disease shared one characteristic - that of wine drinking. However, deaths from road traffic accidents and cirrhosis of the liver tended to be higher than in 'northern' beer-drinking countries. The authors of the 1979 study reported that it would be a sacrilege to attempt to isolate the active principle since it was already available in such a highly palatable formulation! But isolate it somebody did and found dimeric procyanidins which influence cholesterol metabolism. These procyandins are also known as OPC's (oligomeric procyanidin complexes) and pycnogenols.

Many experts ignored this study, stating that the so-called 'French Paradox', i.e. low levels of CHD even though there is a high intake of dairy/saturated fats, was a result of faulty statistics. However, statistics published in 1992 show clearly that this paradox does exist: that levels of heart disease are low in parts of France, even though risk factors, i.e. cholesterol levels, BP, cigarette smoking, are basically the same as in areas where deaths from CHD are high. The paradox is partly due to the use of vegetable fats - particularly olive oil; it is also due to the use of Garlic but is significantly due to red wine drinking. The alcohol itself plays a role since small amounts may protect against CHD, but high levels increase the risk. More and more it appears that the protective factor in red wine is the phenol complex responsible for the red colour: this complex includes catechins, flavonoids, soluble tannins and anthocyanins. These are all known to prevent oxidation, e.g. of LDL-cholesterol to damaging free radicals, and they can also scavenge any free radicals which are formed. Oxidised LDL-cholesterol plays a major role in the development of atherosclerotic plaque. Red wines contain up to 1 g/litre of these phenolics. These key properties have been demonstrated during *in vitro* tests and also in tests in humans given red wine to drink. Wine also contains resveratrol - a phenolic phytoalexin found in grape skins. (Phytoalexins are natural antifungal agents.)

Because grape skins are in longer contact with the juice when red wine is being made, more resveratrol is found in red wine than in white wine. In laboratory tests resveratrol increases the level of the 'good' High Density

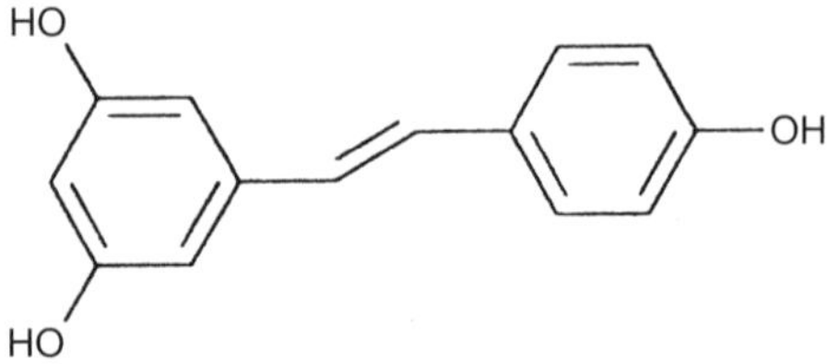

cis-Resveratrol

Lipoproteins and also lowers platelet aggregation. These effects may help to slow the narrowing of coronary arteries through atherosclerosis. Recent research on reservatrol has concentrated on its potential as a chemopreventive agent not just in heart disease, but also in cancer because *in vitro* and *in vivo* tests show significant inhibition of abnormal cell growth. Resveratrol in wine and grape juice also has antiinflammatory effects. There is an increasing consensus that moderate consumption of alcohol and of red wine in particular can reduce the risk of coronary heart disease. For those for whom alcohol is not an option, there are products available which contain procyanidins from grape skins and grape seeds.

Grape seed extract

Animal studies using extracts of the seeds of *Vitis vinifera* have shown beneficial effects relevant to heart disease factors. These extracts generally contain up to 85% of procyanidins. Clinical trials have mainly concentrated on effects in peripheral vascular disease. One trial investigated the antioxidant potential of a commercial extract in healthy volunteers and found that serum antioxidant activity increased significantly for up to three hours after ingestion.

Flavonoids from other sources, e.g. tea, onions and apples, have also been linked with a reduced risk of death from CHD. A study of elderly men in the Netherlands has shown that the higher the intake of dietary flavonoids from these sources, the lower the risk of CHD. If the flavonoid intake was greater than 30 mg per day then the relative disk of death from CHD was less than half that in men who consumed less than 19 mg of dietary flavonoid.

Green tea

It is known that extracts of teas made from the leaves of *Camillia sinensis* have antioxidant effects of relevance not only to heart disease but also in terms of cancer because the antimutagenic activity correlates strongly with the antioxidant activity. Both activities do, however, vary with the type of tea because unfermented teas are more active than fermented black teas. Green or unfermented tea is rapidly steamed or fried after collection to prevent fermentation by inactivation of oxidative enzymes. Green tea contains 30–40% of phenolics such as catechins compared to 3–10% in black tea where the fermentation gives the characteristic aroma of tea and the polyphenols are converted to polymers which contribute to the flavour and colour associated with black tea. A third type of tea called oolong tea is partially fermented.

Commercial extracts of green tea are marketed for the prevention of heart disease and of cancer because of the powerful antioxidant and free radical scavenging properties of the polyphenols. In the case of heart disease it is known that in rats green tea extracts decrease plasma cholesterol and lipid levels. Epidemiological studies in Japan suggest a relationship between lower LDL-cholesterol and increased green tea drinking.

In terms of cancer prevention, animal evidence of benefits from tea is based on *in vitro* studies which show that many reactions involved in tumour proliferation are either suppressed or blocked by tea polyphenols. It is believed that green tea has beneficial effects in reducing certain cancer types in some populations but a direct causal relationship in human trials needs to be established. Given the safety of tea and its potential to exert beneficial effects, such trials will undoubtedly be conducted in the near future.

Garlic and ginseng

The constituents of garlic bulbs, in particular the sulphur derivatives such as alliin, ajoenes, vinyldithines etc., are known for their lipid and cholesterol effects as well as antithrombotic activity. Thus garlic can have a key chemopreventive role in heart disease.

Some have extended the role of garlic by suggesting that it may reduce the risk of certain cancers but the encouraging results obtained in some tests do require confirmation. Another well-known medicinal plant which may have preventive properties against cancer is Korean Ginseng (*Panax ginseng*). Epidemiological studies in Korea have shown that the risk of cancer is significantly reduced in those consuming Ginseng compared to those who do not. The reduction in risk is greater for red Ginseng (produced by steaming the roots) than for white Ginseng.

Dietary fibre and the prevention of cancer

Epidemiological studies indicate that 80–90% of human cancers can be directly linked to factors such as tobacco smoking (which is the single most preventable cause of premature death), alcohol, sunlight, viruses and diet. In the latter case it is known that there is an inverse relationship between high levels of consumption of fruits and vegetables and low levels of cancer. Equally, the consumption of high levels of red meat and saturated fat are an indicator of a high level of cancers. Linked to this is the epidemiological observation that levels of cancer (and indeed coronary heart disease) are lowest in populations, e.g. African, who consume a low fat, high fibre diet.

Dietary fibre consists of all components of plant cell walls not digested by human alimentary enzymes. Chemically, fibre is a mixture of cellulose, lignin, heteropolysaccharides (hemicelluloses), acidic polysaccharides and pectin. In the colon, bacteria hydrolyse up to 15% of cellulose and 70–95% of other polysaccharides giving rise to volatile gases and low molecular weight fatty acids (acetic, butyric etc.) which have weak laxative activity. In addition the cellulosic and non-cellulosic polysaccharides are hydrophilic and absorb large amounts of fluid from the gut lumen, increasing in bulk as they do so. This is exactly the same mechanism that operates with the bulk laxatives such as Ispaghula and Sterculia. The increased bulk arising from the dietary fibre increases stool weight above the minimum needed for defecation. Thus transit time through the bowel is speeded up. This reduces the risk of constipation, diverticulitis, haemorrhoids, varicose veins, hiatus hernia, deep vein thrombosis, acute appendicitis and cancer of the colon. It has been postulated that such cancers may be initiated by exposure to carcinogens produced as a result of bacterial metabolism of bile salts. In a fibre-rich diet such carcinogens would either be absorbed by the fibres or eliminated as a result of the speeded-up transit time. Much of the evidence for the fibre hypothesis is epidemiological, but more recent evidence suggests that other factors may need to be considered because a large study involving over 80,000 nurses over a 16-year period found no association between dietary fibre and risk of colorectal cancer. It may be that it is a combination of mechanical, chemical and pharmacological factors which is involved, because not only do the common sources of fibre such as wheat bran, wholemeal bread, peas, beans, lentils and other pulses provide large amounts of fibre, but many of them also contain antioxidant vitamins and flavonoids as well as compounds known as phytooestrogens.

Phytoestrogens and the prevention of cancer

Phytoestrogens

Phytoestrogens are a family of compounds found in plants which have some oestrogenic and/or anti-oestrogenic activity in animals and/or in humans. These compounds are generally considered to be 'weak' oestrogens and therefore do not participate in female physiology at the same level of potency as do endogenously produced oestrogens, their metabolites or the industrially produced hormones used in therapy. There are two main chemical groups with phytooestrogenic activity, isoflavones and lignans.

Isoflavones

Isoflavones are a unique subgroup of phytooestrogens, but they are also a subgroup of flavonoid compounds. Their precursors, which are abundant in soybeans, are converted to biologically active forms by the action of intestinal bacteria. The key compounds from this group are genistein, daidzein, formononetin and the metabolite equol.

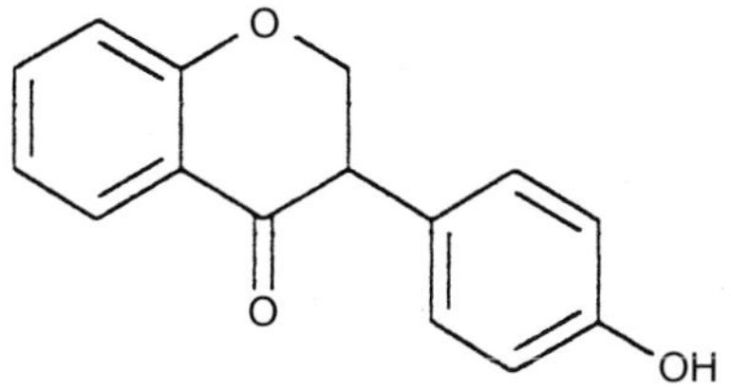

Genistein

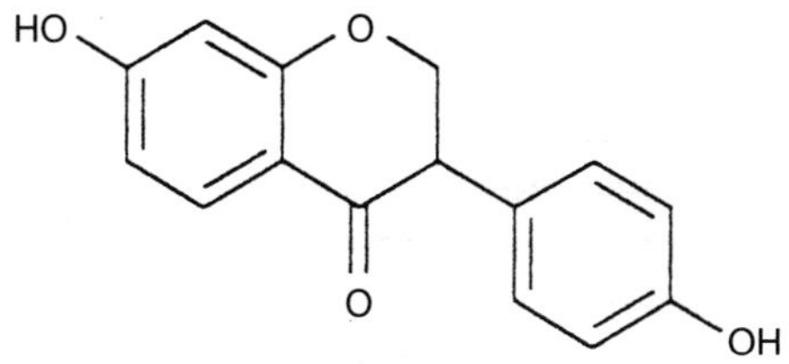

Daidzein

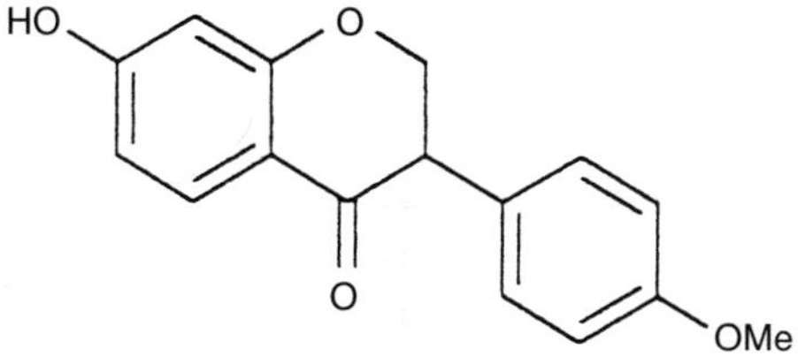

Formononetin

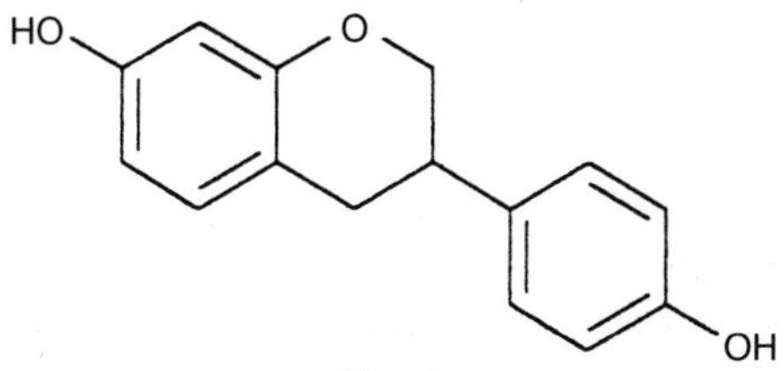

Equol

Lignans

These dimeric phenolic compounds were originally identified in Linseed and are now known to occur in other plant seeds. Lignans have now also been isolated from humans and other animal species. The biologically active forms of these compounds, formed by the action of intestinal bacteria on precursors from dietary grains, also appear to have weak oestrogenic and anti-oestrogenic properties. The first two mammalian lignans to be identified were enterodiol and enterolactone. These are produced by the action of intestinal microflora on the lignan secoisolariciresinol found as a glycoside in linseed.

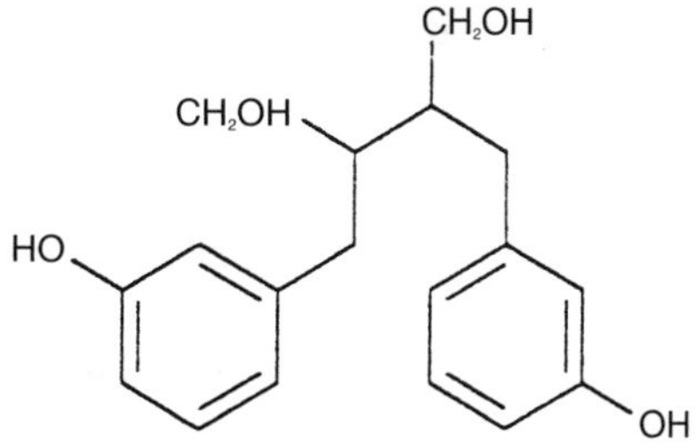

Enterodiol

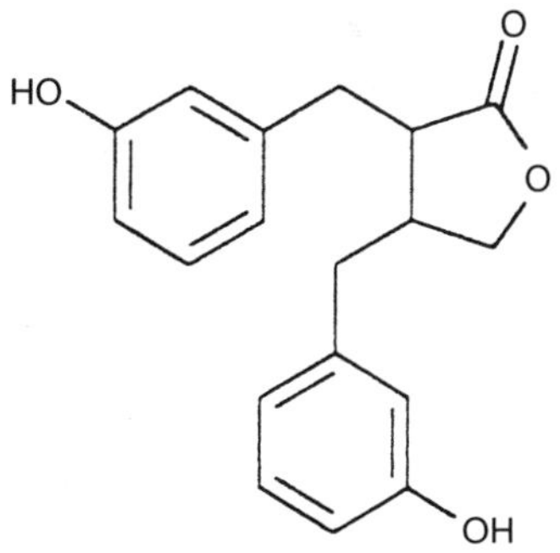

Enterolactone

Phytooestrogenic activity

When studied by a traditional, oestrogen-metabolism bioassay method, genistein was seen to be a very weak oestrogen. Its estrogenic potency was estimated at only 10^{-6} times the potency of diethylstilboestrol.

Isoflavones were shown to act as modulators of endogenous oestrogen activity through their affinity for oestrogen-receptor binding sites. Thus, they could have either oestrogenic and/or anti-oestrogen activities. The latter occurs through competitive inhibition of oestrogen receptors.

Subsequent endocrine research has confirmed that there are a combination of effects on both oestrogen and androgen receptors by phytooestrogen compounds including genistein, daidzein and coumestrol and by the lignan conversion products, enterodiol and enterolactone.

Dr Keith Setchell was the first to identify and report the presence of the weak oestrogen equol in human urine. More than 15 phytooestrogens have now been identified in human urine. Dr Herman Adlercreutz reported that higher levels of enterolactone and enterodiol were found in the urine of vegetarian postmenopausal women than in the urine of omnivorous women. Adlercreutz also showed that these phytooestrogen metabolites appeared in higher amounts in the urine of women who consumed diets known to correlate with low cancer risk compared to women in high-risk areas and in breast cancer patients. This was another piece of evidence suggesting a correlative relationship between hormone metabolism/status in women and dietary patterns and/or constituents.

It is known that Japanese women who maintain a high intake of soy products have a low incidence of breast cancer and few menopausal symptoms. The hypothesis is that phytooestrogens act as natural selective-oestrogen receptor modulators influencing oestrogenic responses in the cardiovascular system, breast and uterus. There is good evidence that eating soy protein lowers cholesterol levels and may help prevent atherosclerosis. Symptoms of the menopause which respond to phytooestrogens include osteoporosis and hot flushes. In the case of osteoporosis it would appear that a daily dose of 90 mg of isoflavones is necessary.

Most attention has focused on whether breast cancer risk can be reduced by phytooestrogen consumption. Three out of four case-control studies (where groups of patients are carefully matched) have indicated that soy intake reduces the risk. One study assayed blood and urine samples from 144 women with newly diagnosed breast cancer and compared their levels of isoflavones and lignans with controls matched for age and residence. It was found that high excretion of both equol (from isoflavonoids) and enterodiol (from lignans) was associated with a substantial reduction in breast cancer risk. In the case of equol there was a four-fold reduction in risk with the highest level of excretion. However, it is still not clear whether it is the phytooestrogens which are the active compounds because they are also markers for a fibre-rich diet since the lignans in particular are found in fibre-rich foods such as whole grains, berries, fruit and vegetables and especially linseed. On the other hand laboratory studies have shown phytooestrogens to have antiproliferative effects on human breast-cancer cells. There is evidence that isoflavone-type compounds may act as antioestrogens by competing with oestradiol for oestrogen-binding sites, thus inhibiting the growth and proliferation of hormone-dependent cells. Both types of phytooestrogens may stimulate Sex-Hormone-Binding Globulin (SHBG) in the liver and thus reduce the amount of free active oestradiol in plasma. Several compounds inhibit the enzyme aromatase, which converts androstenedione to oestrone, and this again may reduce the amount of circulating oestrogen.

Concern has also been expressed concerning the presence of phytooestrogens in soy-based infant formulas but some of the epidemiological studies among Asian women suggest that the protective effect occurs early in life. There is no clear consensus on whether phytooestrogens should not be recommended for women with oestrogen-receptor positive breast cancer.

Cancer of the prostate

The epidemiological studies of Asian populations referred to above also highlight a lower frequency of prostate tumours compared to Western populations. Prostate cancer is the second most frequently diagnosed malignancy in men after skin cancer and is the second leading cause of cancer-related death (after lung cancer).

In Japan mortality from cancer of the prostate is lower than in Western countries. A number of other groups and nationalities also have a low risk of developing this type of cancer, e.g. Seventh Day Adventists who eat a lot of pulses (peas, beans, lentils) had plasma levels of testosterone and 17 β-oestradiol which were significantly lower than those who did not consume diets rich in these compounds as well as having a significantly decreased incidence of prostate cancer. Hawaiians of Japanese ancestry who eat rice and tofu (a soybean product) also have a lower risk of developing this cancer.

One study compared the urinary excretion of isoflavonoids in Japanese men with those of Finnish men and showed that the mean levels for four isoflavonoids were 7–110 times higher in the Japanese men. Two of the flavonoids measured, genistein and equol, are the most oestrogenic of the isoflavonoids and equol is the cause of clover-induced infertility in sheep; this suggests that these compounds could affect oestrogen-sensitive prostate cancer cells. Soy itself is known to be protective against dysplasia of prostatic tissue in mice. The isoflavonoids genistein and biochanin A inhibit androgen-independent prostate cancer cells grown in culture. Genistein inhibits tyrosine-kinases, topoisomerase II and malignant angiogenesis as well as influencing enzymes involved in cell growth and replication. Thus, there is a significant body of biochemical and epidemiological evidence that isoflavonoids lower the risk of prostate cancer. Commercial products containing phytooestrogen-enriched extracts of Soya (*Glycine max*) and of Red Clover (*Trifolium*) are now available.

Conclusion

The combination of biochemical and pharmacological data, together with epidemiological studies, gives strong support to the concept of using phytochemicals of various types as chemopreventive agents. It must be stressed that these compounds and the plants containing them have protective rather than curative roles. The health impact of phytooestrogen-enriched foods or extracts remains to be evaluated for both efficacy and safety. The question of dosages, above those normally present in fruits and vegetables, is one which needs urgent attention.

Reading list

1. *The Food Pharmacy.* Carper J. Simon and Schuster London 1990.
2. *The Phyto Factor* Stewart M. Vermillion London 1998.
3. Adlercreutz H.Does Fiber-rich food containing animal lignan precursors protect against both colon and breast cancer? An extension of the "Fiber Hypothesis". *Gastroenterology* 86: 61-66. 1984.
4. Wattenberg L.W. Inhibition of neoplasia by minor dietary constituents. *Cancer Research* (Suppl.) 43: 24485-24535 1983.
5. Messina M., Barnes S., Setchell K.D. Phyto-oestrogens and Breast Cancer. *The Lancet* 350: 971-972. 1997
6. Aldercreutz H., Markkanen H. and Watanabe S. Plasma concentrations of Phyto-oestrogens in Japanese men. *The Lancet* 342: 1209-1210. 1993
7. Bombardeili E. and Morazzoni P. *Vitis vinifera L. Fitoterapia* LXVI (4) 291-319, 1995.
8. Exploring the power of Phytochemicals: Research Advances on Grape Compounds. Conference Proceedings. *Pharmaceutical Biology* 36 Supplement 1998.Δ
9. St. Leger A.S., Cochrane A.L. Moore T. Factors associated with cardiac mortality in developed countries with particular reference to the consumption of wine. *The Lancet* 1979 (i): 1017-1020.
10. Renaud S. and de Lorgeril M. Wine, Alcohol, Platelets and the French Paradox for Coronary Heart Disease. *The Lancet* 339 1523-1525 1992.

CHAPTER VIII
TOXICOLOGICAL PHARMACOGNOSY

In the foregoing chapters the clinical use of medicinal plants in therapeutic doses has been described. When high doses have been used - deliberately or accidentally - symptoms of intoxication and poisoning of varying degrees of severity may appear. This is because plants contain a number of naturally occurring toxicants, some of which are formed during normal metabolism, some of which are produced as a result of microbial infection, injury or parasite attack. Not only the medicinal plants described earlier but also wild and cultivated plants, trees, bushes and herbs may be poisonous. Some health problems may arise from the non-medical use of plants and some from the use of herbal medicines containing toxins.

A. Poisonous plants

Serious symptoms are seldom seen. The most common symptoms are stomach troubles, stomach pains, vomiting, diarrhoea, that can lead to loss of fluid. More specific problems might occur with certain plant poisonings; for instance dilation of the pupil, redness of the skin, heart palpitations in poisoning with atropine, a component of Belladonna and Datura-species, disturbances in cardiac rhythm (Digitalis, Nerium, Aconite), convulsions (Aconite, Cowbane); these are among the most poisonous species of plants.

A selection of poisonous plants, wild and cultivated, which have caused poisonings in Western Europe follows:

AUTUMN CROCUS (or Meadow-Saffron) *Colchicum autumnale*, L., family Colchicaceae, is cultivated as an ornamental plant in gardens. The whole plant, mainly the seed, contains the alkaloid colchicine.

The symptoms of poisoning start after several hours delay with a burning feeling in the mouth and throat and difficulties in swallowing. Thirst, vomiting and severe and bloody diarrhoea follow. Painful urination, cardiac arrythmia, difficulty in breathing and cyanosis, which causes agony, then occurs. The reflexes in the legs disappear, with loss of sensitivity. The poisoning has a prolonged duration.

BANEBERRY *Actaea spicata*, L., Family Ranunculaceae is a herb found in European forests. When fresh, the whole plant is poisonous. The symptoms of poisoning are similar to those of Meadow Buttercup.

BELLADONNA *Atropa belladonna*, L., Family Solanaceae, is a perennial herb, native to Central and Southern Europe and cultivated in many other countries. Its therapeutic use is due to atropine. Pronounced symptoms of poisoning are seen after children eat 3-5 unripe berries (15-20 for adults).

Mild poisoning gives dryness in the mucosa and skin, thirst, hoarseness and difficulty in swallowing. Severe poisoning gives redness of the skin, rapid pulse and respiration and raised blood pressure. The pupils are widened and do not react to light, vision is blurred. Hallucinations, unrest and restlessness are part of the intoxication.

BITTERSWEET *Solanum dulcamara*, L., Family Solanaceae, The whole plant is poisonous, mainly the stem and leaves. When chewing the stalk of the plant, the taste is firstly acrid then sweet (the reverse of the Latin name of the species).

Symptoms caused by the glycoalkaloid solanine and its aglycone, solanidine, follow the eating of 4–5 berries. The first symptom is a harsh feeling in the mouth and throat. Then comes vomiting and diarrhoea, spasms and fever. After more severe intoxication, which is very rare, there will be blood in the faeces and urine, paralysis and unconsciousness and respiratory paralysis.

BLACK NIGHT SHADE *Solanum nigrum*, L., Solanaceae, is a common weed in gardens and waste places. The whole plant is toxic due to the occurrence of the alkaloid solanine, giving the same symptoms as Bittersweet.

BURNING BUSH *Dictamnus albus*, L., Family Rutaceae, is an ornamental plant. Contact with the aerial parts of the plant gives dermatitis, possibly blisters, that are aggravated by sunshine and increased pigmentation appears. The symptoms can last for weeks.

CHERRY LAUREL *Prunus laurocerasus*, L., Family Amygdalaceae, is cultivated in Western Europe as ornamental shrub. It is mainly the leaves but also buds, bark and seeds which are poisonous due to the occurrence of cyanogenetic glycosides. The symptoms can include delayed nausea, vomiting and palpitation of the heart. In serious cases respiratory depression, unconsciousness and spasms occur.

CHRISTMAS ROSE *Helleborus niger*, L., Family Ranunculaceae, is cultivated in gardens. The whole plant is toxic. Especially in the root there are cardiac glycosides, which are very stable. They are not degraded by drying or storage. The symptoms are the same as for Foxglove.

COMMON BROOM *Cytisus scoparius*, L., Link, Family Fabaceae, is a shrub growing in Europe and Western Asia. The whole aerial part of the shrub, including the seeds, is poisonous due to the presence of the alkaloid sparteine. The sparteine content in the leaves and branch tips is at its highest in the month of May and decreases after flowering, which takes place in June.

Sparteine has an effect on atrioventricular conduction, which resembles that of quinidine, and it also has an oxytoxic effect, i.e. gives a contraction of the uterus. Sparteine has a limited use as oxytoxic agent. Poisoning with Common Broom is unusual, but a lethal poisoning after intake of a decoction has been described. In severe cases there is circulatory collapse and ileus.

COWBANE *Cicuta virosa*, L., Family Apiaceae, grows in shallow waters. All parts of the plant, especially the rhizome and lower part of stalk, are poisonous. The lethal dose for adults is one rhizome. The symptoms might appear within minutes to hours with a burning sensation in the mouth and throat, increased secretion of saliva, stomach-ache, vomiting, diarrhoea and fever.

Cicutoxin is a poison that attacks the central nervous system giving rise to epileptic-type (grand mal-like) convulsions that can be lethal through respiratory paralysis.

CYPRESS SPURGE *Euphorbia cyparissias*, L., Family Euphorbiaceae, is a cultivated ornamental plant found in European gardens. The latex-containing aerial part of the plant is locally irritant. The symptoms of poisoning are irritation of the mucosa in the mouth and throat. In severe cases vomiting, stomach-ache and severe diarrhoea occur. Later on there is mental confusion, circulatory disturbances, cold sweat, and dilatation of the pupil. Other *Euphorbia* species also contain latex with an irritating effect on mucosa and skin.

FOOLS PARSLEY *Aethusa cynapium*, L., Family Apiaceae, is a European weed. The whole plant is poisonous due to the occurrence of very low content of the alkaloid coniine, a cicutoxin-like substance and essential oil. There have been lethal poisonings among humans and cattle. Symptoms of poisoning are stomach-ache, increased production of saliva, vomiting, diarrhoea, headache and blurred vision.

Typically there is the progressive paralysis of respiration and the intact consciousness.

FOXGLOVE *Digitalis purpurea*, L., Family Scrophulariaceae, is a decorative garden plant. The leaves are poisonous due to its content of cardiac glycosides: 2–3 g dried or 12–18 g fresh leaves, which have an unpleasant sharp taste, give the pronounced poisoning symptoms. The mucosa in the mouth and stomach will become irritated if the leaves are placed in the mouth or eaten. After several hours comes nausea, vomiting, headache, tiredness and drowsiness followed by disturbance in the cardiac activity (arrythmia) and possibly spasms.

HENBANE *Hyoscyamus niger*, L., Family Solanaceae, grows on cultivated farmland. The whole plant is poisonous, mainly the seeds due to the presence of L-hyoscyamine, atropine and scopolamine. The symptoms of poisoning are the same as for Belladonna.

HERB PARIS *Paris quadrifolia*, L., Family Liliaceae, has berries that can be mistaken for blueberries. Severe poisoning occurs after consuming 30–40 berries. The symptoms start with irritation of the mucosa, nausea, vomiting, stomach-ache, and painful spasms. Other symptoms which might occur are headache, dizziness and miosis.

HOGWEEDS *Heracleum sibiricum*, L., *H. mantegazzianum* L., and *H. spondylium*, (Family, Apiaceae), are decorative plants, also considered as weeds. Contact with fresh leaves and stems causes severe dermatitis, which is aggravated by ultraviolet light: liquid-filled blisters may occur that might turn into purulent wounds. Scarring of the skin is common, as is hyperpigmentation. The skin reaction varies with the individual; many cases of strong local reactions are known among children and adults. This is known as phytophotodermatitis and is due to the presence of furocoumarins (psoralens) in the sap. Not all furocoumarins are phototoxic. Their potential to cause skin problems varies depending on whether the molecule is linear (strongly phototoxic) and on the substituents of the molecule. Severe reactions are associated with psoralen, 8-methoxy psoralen (xanthotoxin), 5-methoxy psoralen (bergapten) and 5,6-dimethoxy isopsoralen (pimpinellin). Sources of these compounds other than the Giant Hogweeds, which present the main threat to humans, include Bergamot lime (*Citrus bergamia*), Celery (infected) (*Apium graveolens*), Parsnip (*Pastinaca sativa*), Rue (*Ruta graveolens*) and *Dictamnus* already mentioned. In terms of garden plants Rue is the main cause of problems while escapes of Heracleum species present the major threat in the wild.

LABURNUM *Laburnum anagyroides*, Medikus and *Laburnum alpinum* (Miller) Bercht et J. Presl., Family Fabaceae, are well-known ornamental trees in the garden. In spring they produce yellow flowers in large hanging racemes and the fruit is a capsule similar to a peapod.

The poisonous part of the plant is above all the seeds due to the occurrence of the alkaloid cytisine; 2–10 seeds can give symptoms of poisoning in children as can 3–4 pods for an adult. Within 15 minutes to a few hours the symptoms start: the mucosa of the mouth becomes irritated and the secretion of saliva increases. Then follows violent and continuous, sometimes bloody, vomiting and pain in the pit of the stomach.

Diarrhoea and fever, pupillary dilation, thirst, cold sweats and headache might occur.

Agony and hallucinations might occur. Pulse and respiration are more rapid, then more superficial.

LILY-OF-THE-VALLEY *Convallaria majalis*, L., Family Convallariaceae, contains about 20 cardiac glycosides; the highest concentration occurs in flowers and berries, less in leaves and rhizome. Children have been intoxicated when they have eaten the red berries. After a latent period the symptoms are nausea, vomiting, diarrhoea and possibly cardiac irregularities. The suggested use of Lily of the Valley flowers in salads is highly dangerous.

MEADOW BUTTERCUP *Ranunculus acris*, L., Family Ranunculaceae, is a well-known plant. The whole plant is poisonous when fresh. Consumption of fresh plants gives a burning feeling in the mouth and throat caused by protoanemonin. Then follows, vomiting, stomach-ache and bloody diarrhoea. Dizziness, spasms, and irritations of the urinary tract with decreasing painful dysuria are among other symptoms. In very severe cases there is circulatory and respiratory collapse.

MEZERON *Daphne mezereum*, L., Thymelaceae, is cultivated as an ornamental plant. All parts of the plant are poisonous, mainly the bark and the berries, with a sharp taste, through the occurrence of mezerine.

10–15 berries is a lethal dose for adults. The poisoning starts with a burning feeling in the mouth and throat, difficulties in swallowing, stomach-ache and bloody diarrhoea followed by pallor and cold sweats, respiratory and circulatory collapse.

MONK'S HOOD (Wolfsbane), *Aconitum napellus*, L., Family Ranunculaceae, grows wild in Europe and is cultivated as an ornamental plant. The whole plant, mainly the root and seed, is poisonous through the occurrence of the alkaloid aconitine (0.5–1.5% in the root). The aconitine content of the drug varies considerably because the alkaloid is easily hydrolysed to the weakly active compounds benzoyl-aconine and aconine.

The symptoms start with a burning feeling in the mouth and throat, followed by increased secretion of saliva, vomiting and possibly diarrhoea, disturbances of consciousness, thirst, perspiration and disturbances in vision. A progressive paralysis is typical in aconitine intoxication, which can lead to respiratory collapse and/or ventricular fibrillation at the very least.

POISON HEMLOCK *Conium maculatum*, L., Family Apiaceae The whole plant, including the fruits, is poisonous through the occurrence of coniine (90%) and related alkaloids (10%). Poisoning with poison 'hemlock' nowadays is due to confusion with other plants of the kitchen garden. Contamination of aniseed with the similar-looking 'hemlock' fruits is an example. Unripe fruits have a higher alkaloid content (2%) than other parts of the plant. On ripening the content decreases to about 0.7%. Coniine is relatively unstable and therefore the content in stored materials is low. In the case of fresh leaves and flowers the lethal dose for an adult is 6–8 'hemlock' leaves. The symptoms start with increased secretion of saliva; then follows vomiting, thirst, disturbances of sight, muscle weakness and a progressive paralysis of respiration. However, consciousness remains intact to the very last.

SAVIN *Juniperus sabina*, L., Family Cupressaceae, is cultivated as an ornamental shrub. Severe poisoning has occurred when the plant has been used as an abortifacient due to its content of sabinol, which irritates the mucosa of the gastrointestinal tract and leads to stomach-ache, possibly bloody diarrhoea, and disturbed activity of the kidney. 5–20 g of the drug gives spasms, loss of consciousness and possibly respiratory collapse.

THORNAPPLE *Datura stramonium*, L., Family *Solanaceae* is an annual herb, occurring in Europe, North and South America. Occurrence of alkaloids and poisoning symptoms are the same as for Belladonna.

WATER DROPWORT *Oenanthe aquatica*, L., Poir., Family Apiaceae, grows in ponds and small lakes in Europe. The flowers are the most poisonous, followed by root, fruit and stem. Phellandrene, which has an abortive action, gives first a stimulating effect and then inhibition of respiration. Oenanthin provokes vomiting and oenanthotoxin is a poison that gives epilepsy-like convulsions.

WOOD ANEMONE *Anemone nemorosa*, L., Family Ranunculaceae, is a well-known herb, flowering in spring. The whole plant in fresh condition is poisonous through the occurrence of protoanemonin, which is also found in Buttercup. Therefore the symptoms of poisoning are the same.

YEW *Taxus baccata*, L., Family Taxaceae, is mainly cultivated as a decorative shrub in gardens. The whole plant, with the exception of the red 'fruit', is poisonous due to the alkaloid taxine. Children who have eaten the red berries and chew the taxine-containing seeds and needles can be poisoned. However, if the seeds are swallowed intact, the taxine will not be absorbed. Vomiting, stomach pains and diarrhoea are initial symptoms, which appear within half to one hour; further symptoms include cardiac arrythmia, weakness of the muscles and possibly circulatory collapse.

B. Plants as drugs of abuse

The non-medical use of drugs has become a significant health and social problem over the past twenty years. Problem drug-taking is a complex phenomenon involving human beings, the drug or drugs they take and the social economic and cultural environment in which the drug-taking takes place. All these aspects are complex and all can contribute to what is usually referred to as the 'drug problem'. The drugs themselves play a key part because of the manner in which they bind to key neurotransmitter receptors. All of the drugs are psychoactive or mood altering and most are either plant materials, secondary metabolites or compounds easily derived from natural products. While there is increasing evidence that drugs of abuse affect many of the same neurotransmitter systems, it is useful to catalogue the different drugs into separate pharmacological categories such as:

1. Depressants
2. Stimulants
3. Narcotics
4. Hallucinogens

The Depressant group is represented by ethanol in its various forms produced through the fermentation of starches and sugars by yeast. Under the heading Stimulants can be included cathinone, which is structurally similar to amphetamine ('speed') and which occurs in the leaves of *Catha edulis* ('Khat' or 'Qat'). Another stimulant is the alkaloid cocaine from Coca (*Erythroxylon coca*) leaves which is inhaled or snorted through the nose as a 'line' of cocaine hydrochloride or smoked in the form of the basic alkaloid so-called 'crack' or 'free-base'. Nicotine found in tobacco (*Nicotiana tabacum*) is a drug which can have either stimulant or calming effects depending on the user's mood. Similarly the effects of the various Cannabis drugs can be difficult to categorise pharmacologically, whether one is referring to the herbal form ('Marijuana', 'Grass', 'Weed', 'Skunk'), to the resinous 'Hashish' or 'Hash' or to the distilled 'Hash Oil' or 'Liquid Hash', because of the variety of effects which can be caused by the complex phytochemical mixture found in *Cannabis sativa*.

The Narcotic analgesics from the Opium Poppy (*Papaver somniferum*) i.e. morphine and the semisynthetic derivative heroin

(Diacetylmorphine), constitute a major group. These opiates cause intense euphoria when they are injected or smoked. Repeated use of morphine or heroin almost invariably leads to physical and psychological dependence. Toxicologically the main effects of an opiate such as heroin are caused by the fact that the drug is injected, leading to serious infections such as HIV and Hepatitis. Opiate dependence is associated with premature death through overdoses, disease, crime and general alienation from society.

Hallucinogens

These drugs, also referred to as psychotomimetics, psychedelics or deliriants, can be defined as 'plants or chemicals which, in non-toxic doses, produce changes in perception, thought and mood with-out causing major disturbance of the Autonomic Nervous System'. Most hallucinogens are of plant or fungal origin. Some, e.g. LSD, known as 'acid' after its chemical name Lysergic Acid Diethylamide, is semisynthetically produced from the ergot alkaloids. The drug 'Ecstasy' (methylendioxymethamphetamine) is synthetic but is synthesised from safrole or isosafrole found in a number of plant essential oils, e.g. *Sassafras albidum*. A significant feature of the 'Ecstasy'-type drugs is the methylenedioxy group which also occurs in myristicin, found in nutmeg (*Myristica fragrans*), and probably explains the hallucinatory effects experienced when large amounts of nutmeg are consumed.

In general the chemicals responsible for hallucinogenic effects are confined to a relatively small number of chemical classes:

1. ***Tropane Alkaloids*** found in plants of the family Solanaceae, especially *Datura suaveolens, Brugsmansia* and *Brunsfelsia* species. The alkaloid Hyoscine (scopolamine) in high doses can cause hallucinations giving the sensation of flying through the air, hence the possible link between medieval witchcraft and the use of Deadly Nightshade (*Atropa belladonna*) which contains similar alkaloids. These alkaloids and the plants containing them are known to be poisonous and fatalities have been reported.

2. ***Phenethylamine Alkaloids*** are found in various Cacti, particularly Echinocacti such as *Lophophora williamsii* known as 'Peyote', 'Peyotl' or 'Mescal Buttons'. The main alkaloid is mescaline. Peyote is used for religious purposes by Native Americans because the hallucinations and alterations in perception are believed to be a form of communion with God. Most plant and fungal hallucinogens have a very long history of religious use by aboriginal tribes and cultures, particularly in South America. While modern neuropharmacology can explain the visual and auditory hallucinations as being drug-induced, primitive tribes could only rationalise the effects as being due to the appearance of various deities. Some such plants would be known as 'Plants of the Gods' and their use extends back in some cases to 500 BC.

3. ***Indole Alkaloids*** and the organisms containing them constitute the most important and powerful group of hallucinogens. Some are derived from Lysergic acid, e.g. those with LSD-type effects found in the seeds of two members of the Convolvulaceae: *Ipomoea violacea* and *Rivea corymbosa* known as the Morning Glories or as 'Ololiuqui' used traditionally in central Mexico.

OLOLIUQUI Indian tribes in Mexico use the seeds from two climbers: *Rivea corymbosa* (L.) and *Ipomoea violacea*, family Convolvulaceae. Albert Hofmann, the discoverer of LSD, has also isolated from this drug the hallucinogenic ingredient, lysergic acid amide, with an effective dose of 2–100 mg. The related compound LSD (lsysergic acid diethylamide) is a semi-synthetic substance, one part of which, lysergic acid, is the characteristic part of the ergot alkaloids. Dr Hofmann worked during the 1940s with synthetic amides of lysergic acid and prepared a series, one of which was the diethylamide with absolutely unexpected hallucinogenic effects. He has described this in his laboratory journal for 19 April 1943. The effective dose is only 25–100 micrograms.

TEONANACTL A second major group of indole hallucinogens are based on the tryptamine nucleus and they are found in approximately ten genera of mushrooms, the most important of which are the *Psilocybe* species. These mushrooms were originally used in Aztec rituals and were known as 'Teonanacatl' or 'Flesh of the Gods' but nowadays are more usually referred to as 'magic mushrooms'. In some mountainous regions of Southern Mexico the Indians still use mushrooms for magic ceremonies. These mushrooms belong to the genus *Psilocybe*, above all *Psilocybe mexicana*, Heim (Agaricaceae). The active alkaloids psilocybin and psilocin are characterised by their 4-hydroxy tryptamine structure. In Europe it is *P. semilanceata* that is likely to be used as 'magic mushrooms'.

The mushroom Fly Agaric (*Amanita muscaria*) also has hallucinogenic properties but contains a different compound called muscimol.

The use of the fly agaric, *Amanita muscaria* (L.) Pers. (Fr.) Hook, as an intoxicant by the East Siberian Koriak tribe was first described in 1730 by the Swedish officer Philip Johan Strahlenberg, who was a prisoner of war in Siberia. After eating 1–4 mushrooms – fresh or partly dry – the effects appear after 15–60 minutes, consisting of unrest, euphoria, a feeling of weightlessness and coloured hallucinations. Higher doses (5–10 mushrooms) give more pronounced poisoning symptoms with spasms and minor vivid hallucinations, a delirium-like state, then a narcotic stage, which in rare cases can go into coma and be lethal. The Koriaks also know that the urine from an intoxicated person has a hallucinogenic effect – therefore the urine was also consumed. Among the active substances in the fly agaric, muscarine has a pronounced cholinergic, but not hallucinogenic, effect. Since its content in fly agaric is very low (2–3 mg/kg fresh mushroom), it plays a minor role in the overall effect. The hallucinogenic effect is given by the labile ibotenic acid and its decarboxylation product, muscimol, which is 5–10 times stronger than ibotenic acid. The effective dose of muscimol is 10–15 mg.

The toxic effects of these drugs lies in their behavioural effects rather than any direct poisonous effect (with the exception of the Solanceous plants). Cases of delusional behaviour leading to accidents, frightening hallucinations leading to severe anxiety and paranoia ('bad trips') and prolonged psychotic episodes have all been reported following consumption of these materials.

4. Non-nitrogenous hallucinogens include Nutmeg, myristicae semen. The drug is the dried seed kernel of *Myristica fragrans*, Houtt., family Myristicaceae, a dioecious tree in South East Asia, where it is cultivated (Indonesia, Malaysia, Sri Lanka). (Fig. 32) Since it is a dioecious species, the male trees are kept to about 10% of the total amount of cultivated trees. The yellow, pear-shaped fruit is opened on ripening, and the black seed, surrounded by a red aril, is disclosed. The aril is dried to give the drug Mace. On drying of the seed (in an oven) the kernel shrinks and is freed, when the shell is crushed, to constitute the drug Nutmeg. In order to protect against insect attack the drug is often dipped in a slurry of lime.

In low doses nutmeg, which contains 30–40% fat and 10–15% essential oil, is an excellent spice. The main ingredients of the essential oil are terpenes and also the substances myristicin and elemicin. Both substances are pharmacologically inactive, but in the human body they are partly transformed into amphetamine-derivatives, which have a hallucinogenic action. This

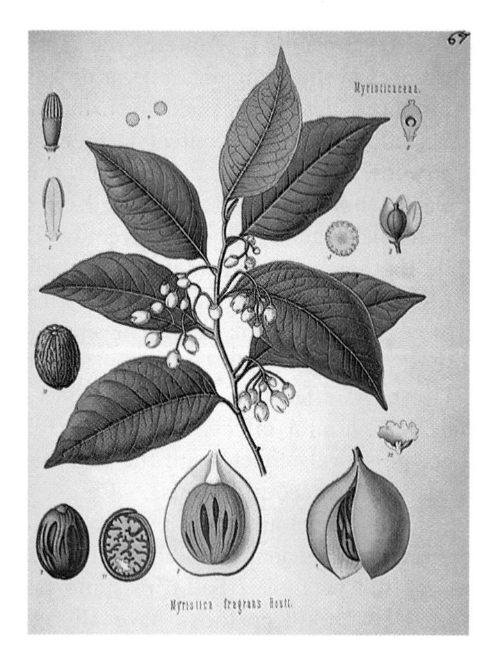

Fig. 32. *Myristica fragrans* Houtt. Flowering branch, fruit with arillus

Muscarine

Muscimole

Mescaline

Cathinone

Arecoline

Psilocybin

action is seen after intake of large amounts of nutmeg.

The long latency in nutmeg poisoning is explained by the biotransformation of myristicin and elemicin which is time-consuming. The hallucinogenic effect appears at the earliest 2 hours after the intake and lasts 12–14 hours. Typical is the loss of sense of time and space and a feeling of floating. This is followed by unpleasant side-effects: reddening of the skin, palpitations and decreased salivation and thirstiness.

KHAT OR QAT *Catha edulis* (Vahl) Forsk. Ex. Endl., family Celastraceae, grows in East Africa. It is cultivated in the mountainous regions (above 1,500 m) in Ethiopia, and Kenya. In countries such as Yemen, the fresh leaves are chewed or used for preparing tea, which is used as a stimulant. Khat counteracts fatigue, facilitates strenuous muscular work, and causes talkativeness. These effects depend on the occurrence in the fresh leaves of the labile alkaloid cathinone, which quickly breaks down into pseudoephedrine. The effects of cathinone are similar to those of amphetamine and there are reports of amphetamine-type dependence and psychotic reactions in Khat users.

CANNABIS *Cannabis sativae herba*, also called Hemp or Indian Hemp, consists of the dried, aerial parts of *Cannabis sativa* L., family Cannabaceae. The plant is an annual dioecious herb, (2–4 m) and is cultivated in temperate and subtropical regions for production of fibres, seeds and illegally for hashish or marijuana. (Fig. 33) The leaves and bracts of both male and female plants have glandular hairs, which secrete a resin rich in cannabinoids, which has euphoriant effects. The female plant is mainly used, since it is bigger and produces a larger amount of resin. Some 60 cannabinoids have been isolated. They are terpenophenolics and the content varies depending on the chemotype, plant part and production conditions. Major cannabinoids include Tetrahydrocannabinol which predominates in the drug type used to prepare marijuana (up to 20%), resin or distilled oil (up to 60%). Cannabidiol (CBD) predomi-

Fig. 33. *Cannabis sativa* L. Cultivation in Turkey. Prof. em. Turhan Baytop, Istanbul in the front.

nates in fibre-type cannabis (Hemp) and has depressant effects in contrast to the euphoriant properties of THC. Cannabinol (CBN) is a decomposition product with very low levels of psychoactivity. Δ^9 - THC (tetrahydrocannabinol) and its main metabolite, 11-OH-Δ^9-THC are the main substances responsible for the intoxication obtained after smoking, the pulmonary absorption being greater than the oral one: about 20% of the Δ^9-THC in a cigarette is absorbed. The effects of cannabis smoking include increased awareness and perception, euphoria, sedation and occasionally hallucinations.

Cannabis has been used as an analgesic and Δ^9-THC is a potent antiemetic; and the drug has also been used as a remedy for glaucoma and in multiple sclerosis. The side-effects have, however, limited the drug in its therapeutic use which are now being intensely investigated.

The formula shows Δ^9-THC and 11-hydroxy Δ^9-THC.

Delta^9 Tetrahydrocannabinol

11-Hydroxy-Delta^9 Tetrahydrocannabinol

Cannabis is a controversial drug. Recent reviews of its effect on human health have highlighted issues such as effects on learning and short-term memory and on reactions and co-ordination leading to accidents. According to WHO, chronic use may lead to a cannabis dependence syndrome and schizophrenia may be exacerbated in affected individuals. Because of the composition of cannabis smoke, which contains carbon monoxide and carcinogens, cannabis smokers have a higher prevalence of acute and chronic bronchitis than non-smokers. There is also evidence of injury to the trachea and bronchi, lung inflammation and impaired defence against infection. There appears to be an increased risk of certain types of cancer either in smokers or in the offspring of mothers who smoke during pregnancy. While cannabis has been used for thousands of years, mass use of the drug by young people is a much more recent phenomenon and further research is needed to clarify the health risks of this complex drug.

Herbal substitutes for drugs of abuse A variety of herbal mixtures are offered for sale in magazines, on the internet and in so-called 'smart', 'eco' or 'head' shops. Many are marketed as herbal 'Ecstasy' and the plants included in the formulations include Yohimbe bark, Kava-Kava (*Piper methysticum*), Valerian, Hops, Jaborandi and Alisma. One product contains Kava-Kava, Guarana, Uva Ursi and Cascara bark. Many of the products sold as herbal 'Ecstasy' contain either *Ephedra sinica* (Ma huang) or the Indian plant *Sida cordifolia* which both contain the alkaloid ephedrine (see RO3c, Chapter VI). Other alkaloids may also occur, such as pseudoephedrine, norephedrine and norpseudoephedrine. The side-effects of ephedrine include tachycardia, anxiety, insomnia and arrythmias and a hypotensive crisis may develop if monamine oxidase inhibitors are also taken. Many adverse reactions and more than 20 deaths have been attributed to ephedrine and *Ephedra* consumption. Research conducted in the US shows that the daily intake of some *Ephedra* products would give ephedrine levels well above the recommended therapeutic doses.

The advertising of *Ephedra* products for both their euphoria-inducing and slimming effects, the publicity surrounding deaths following their use and the resulting attempts to regulate and control the availability of these products raise a number of questions. One question which is frequently asked is that concerning the relative safety of *Ephedra* versus isolated alkaloids. When *Ephedra* is used medically the dosages of the herb used are likely to present less of a hazard than when the drug is used non-medically or when it has been 'spiked' with added synthetic alkaloids, e.g. norpseudoephedrine. In that situation, lives are at risk and athletes who consume such products are at risk of failing drug tests.

A further aspect of concern about ephedrine and *Ephedra* is the fact that it can be used as a precursor for the synthesis of methamphetamine; accordingly, under the UN Convention (1988) against Illicit Traffic in Narcotic Drugs and Psychotropic Substances, the supply of substances such as ephedrine is controlled. Within the European Union, Directive 92/109/EEC places precursors, including ephedrine as well as the essential oil components safrole, isosafrole and piperonal, under surveillance measures.

'Smart' drugs A range of plant materials are sold as 'smart' drugs or drinks. Many contain caffeine in its purified form or in the form of Guarana. *Ginkgo biloba* extracts are very popular, as are Ginsengs of various types. Two plants which are frequently encountered are Kava-Kava (*Piper methysticum*) and Yohimbe (*Pausinystalia yohimbe*) (K. Schum.) Pierre ex Beille. Yohimbine contains up to 5.9% of alkaloids in the stem bark, chiefly Yohimbine. Yohimbine is traditionally used as an aphrodisiac and stimulant. Yohimbine is an α-2-adrenoceptor antagonist. It affects the cardiovascular system and is contraindicated in patients on tricyclic antidepressants and sympathomimetics such as amphetamine, ephedrine and cocaine. Individuals with heart disease, hypertension, renal disease, depression or other psychiatric illness are also at risk. Health professionals need to be aware of the types of plants being sold in this fashion so as to provide objective advice on their health effects.

C Mutagenic and carcinogenic compounds in medicinal plants

Medicines derived from plants are not only expected to be effective and of good quality but also must be safe. Increasing attention is being paid to the chronic or long-term toxicity of herbal medicines. As part of that testing within the regulatory guidelines for toxicity of medicines, many plants have been subjected to tests for mutagenicity and carcinogenicity.

Mutagenicity

If a chemical, whether it be synthetic or natural, reacts with DNA within the cell nucleus, it is considered to be mutagenic and as such able to induce a stable, heritable change in a DNA sequence of any cell or organism coming in contact with the compound. If this mutation occurs in germ cells then the consequence is teratogenicity. If it occurs in somatic cells then carcinogenesis may result. Testing for carcinogenicity is time-consuming (minimum of 2 years), involves large numbers of animals and hence is expensive. A number of *in vitro* tests have been introduced involving bacterial cells, cultured mammalian cells (rat bone marrow cells, human leukocytes or Chinese hamster fibroblasts). Some tests, e.g. the micronucleus test, are *in vivo* tests. The most popular is the Ames test, more correctly known as the Bacterial Auxotrophy Reversion test. This test involving mutant strains of *Salmonella typhymurium* has been used to test literally thousands of natural and synthetic compounds. Substances which give positive results can then be subjected to more detailed genotoxicity and carcinogenicity screening because Ames was able to show that the majority of mammalian carcinogens were mutagenic in bacteria.

Many phytochemicals are mutagenic, leading to concerns about their long-term consumption in phytomedicines and in the diet. One of the most mutagenic compounds in nature is the flavonoid quercetin.

Quercetin occurs in many medicinal plants, for example Hypericum and Crataegus, as well as in fruits and vegetables. It is estimated that the average daily consumption in the diet is 25 mg. Because of its ubiquitous occurrence it has been studied extensively to ascertain if the strong mutagenicity has relevance to human health. *In vivo* studies have generally yielded negative results. A 2-year study in rats investigated possible carcinogenicity and found an increased risk of renal tumours only at the highest dosages and only in males. The general assessment is that the rodent studies have no relevance for humans.

Some plants contain both mutagenic and antimutagenic components, e.g. Ginger (*Zingiber officinale*, Roscoe). In others the mutagenicity does predict subsequent carcinogenicity. One well-known example is *Acorus calamus* (Sweet Flag) where the presence of β-asarone in certain chemotypes was responsible for the mutagenic effects of extracts. These extracts subsequently caused intestinal tumours in rats. As a result the use of Sweet Flag is not recommended.

A variety of phytochemical types are known to be carcinogenic. Included among them are cycasin, a methylazoxymethanol glucoside from cycads consumed by Pacific Islanders, which causes hepatic, intestinal and renal tumours. Bracken fern (*Pteridium aquilinum*) contains a nor-sesquiterpene glucoside, ptaquiloside, which induces adenomas and adencarcinomas of the ileus. The phenylpropanoid compounds safrole, isosafraole and estragole have also come under suspicion. Safrole is found in Sassafras Oil (*Sassafrass albidum*) among other volatile oil-containing spices. It is carcinogenic and hepatotoxic. Since it can be used to synthesise the illegal drug MDMA ('ecstasy') it is also controlled under Drug Precursor legislation.

The most well documented of the medicinal plant carcinogens are the aristolochic acids from various *Aristolochia* species, e.g. *A. clematis* (Birthwort), as well as *Asarum canadense* (Canada Snake Root). Aristolochic acids I and II are nitrophenanthrene carboxylic acids. Both are extremely potent mutagens and carcinogens and are also genotoxic and nephrotoxic in animal models. One

Chinese species (*A. fangchi*) has been implicated in the development of severe kidney disease, requiring dialysis and transplantation, in a group of over 70 Belgian women who were supplied this plant in error, as part of a slimming regime. This problem arose because the plant was purchased using the Chinese common name 'Fang Ji' rather than the Latin binomial name *Stephania tetrandra* (also known along with 21 other plants as 'Fang Ji' in Chinese). Many of the women affected have subsequently developed malignancies. While *Aristolochia* species are no longer recommended for human consumption in the West, they do have a long history of use in herbal medicine and it is possible that the Belgian tragedy involves not only the aristolochic acids, but also an interaction with some of the synthetic drugs included in the formulation.

D Hepatotoxic plants

Plants containing pyrrolizidine alkaloids are both carcinogenic and hepatotoxic. These effects are found in both humans and farm animals. A number of plants contain such alkaloids, such as Senecio species (Ragworts) including *Senecio longilobus*, Coltsfoot (*Tussilago farfara*), Butterbur (*Petasites japonicum*), *Crotalaria*, *Heliotropium* Houndstongue (*Cynoglossum*), Borage (*Borago*) and Comfrey (*Symphytum*) species. The pyrrolizidine alkaloids in these plants are esters composed of a necine base and a necic acid. Both saturated and unsaturated forms occur. Alkaloids with a saturated necine base are non-toxic, e.g. those of *Arnica* and *Echinacea*. Those with an unsaturated base, e.g. senecionine, senkirkine and sympthytine, are hepatotoxic and carcinogenic. Pyrrole derivatives produced by metabolism are believed to be responsible for most of the toxic effects. Consumption of plants containing these alkaloids as food or in herbal remedies, e.g. Comfrey tea or bush tea, results in the development of veno-occlusive lesions in the liver (Budd-Chiari Syndrome) which can progress to liver cirrhosis. Large-scale poisonings due to the consumption of *Senecio* (South Africa) and *Heliotropium* (Afghanistan) contaminated flours have been reported. Serious effects including fatalities have been reported following consumption of herbal teas containing these alkaloids. Many health authorities restrict the availability of herbal medicinal products which produce exposure to unsaturated pyrrolizidine alkaloids above 1 mg internally or 100 mg externally per day. Herbal medicines providing up to 1 mg per day may only be used for a maximum of 6 weeks in any year. A number of other plants and phytochemicals are known to be hepatotoxic, including safrole as already mentioned. Liver damage has also been reported following consumption of *Larrea tridentata* (Chaparral or Creosote Bush). The principal constituent is nordihydroguaiaretic acid which is a powerful antioxidant once investigated for food use and for its potential anticancer effects. Long-term feeding of this compound to rats led to lymph node lesions and *Larrea* is not recognised as a safe plant. Wall Germander (*Teucrium chamaedrys* L.) has been implicated in cases of acute hepatitis, including one fatality when used in a slimming product in France. The diterpenes appear to be the compounds responsible. There is also evidence to suggest that *Teucrium* is sometimes confused with the herb Skullcap (*Scutellaria*) and it is believed that this substitution may explain cases of liver disease reported following consumption of products supposedly containing Skullcap.

Kava-kava (*Piper methysticum*) has recently been implicated in a number of cases of jaundice, indicating possible hepatotoxicity.

E Allergens in plants

In addition to the natural toxins already described, plants also contain allergenic compounds. Many of these are harmful if the plant allergen is ingested or inhaled, though some are contact allergens.

E.1 Ingested allergens These are mainly proteins consumed in foods. The most important are the 37 different allergenic compounds in peanuts (*Arachis hypogaea*). The consumption of peanuts causes deaths through anaphylactic shock each year. Other nuts with allergenic effects include Brazil nuts, Almonds and Pistachio nuts.

E.2 Inhaled allergens Exposure to these commonly results in hay fever and asthma. The main sources are pollens and fungal (mould) spores. Pollens from grasses such as Timothy (*Phleum pratensis*) and perennial Rye (*Lolium perenne*) as well as those of *Plantago* and *Ambrosia* species are important in causing seasonal hay fever and possibly asthma. Spores from fungi such as *Cladosporium* and *Sporobolomyces* have been implicated in rhinitis and asthma.

E.3 Contact allergens The sesquiterpene lactones of the Asteraceae, e.g. chrysanthemums, asters and daisies, are well-known causes of contact allergic dermatitis. Patients with such allergies should avoid (and should be advised to avoid) herbs such as Artemisia (Mugwort), Chamomile, Yarrow (*Achillea*). Feverfew, *Echinacea* and *Arnica* because of cross-sensitisation.

Reading list

1. *Poisonous Plants in Britain and their Effects on Animals and Man*. Cooper M.R. and Johnson A.W. HMSO London 1984.
2. *A Colour Atlas of Poisonous Plants – A Handbook for Pharmacists, Doctors, Toxicologists and Biologists.* Frohne D. and Pfänder H.J. Wolfe Science London 1984.
3. *Botanical Safety Handbook.* Edited by McGuffin M., Hobbs C., Upton R. and Goldberg A. CRC Press Boca Raton 1997.
4. *Essential Oil Safety* Tisserand R. and Balacs T. Churchill Livingstone Edinburgh 1995.
5. *Botanical Dermatology.* Mitchell J. and Rook A. Greengrass Vancouver 1979.
6. Vanherweghem J-L *et al.* Rapidly Progressive Interstitial Renal Fibrosis in Young Women: Association with Slimming Regimen including Chinese Herbs. *The Lancet*, 341: 387–391. 1993.
7. *Adverse Effects of Herbal Drugs.* Edited by deSmet P.A.G.M., Keller K., Hänsel R. and Chandler R.F. Springer Verlag Berlin Volume 1 1992, Volume 2 1993, Volume 3 1997.
8. *Drugs of Abuse.* Wills S. Pharmaceutical Press London 1997.
9. Hall W. and Solowij N. Adverse Effects of Cannabis *The Lancet* 352: 1611–1616. 1998.
10. *Therapeutic Uses of Cannabis.* British Medical Association, Harwood Academic Publishers. The Netherlands 1997.
11. Ashton C.H. Biomedical Benefits of Cannabinoids. *Addiction Biology* 4: 111–126. 1999.

Chapter IX
GLOSSARY

A. Glossary of botanical and chemical terms

Acetogenin: consists of fatty acids, fats, waxes, flavonoids, antraquinones, phloroglucinol.

Achene: fruit, often winged, whose covering does not open at maturity.

Alkaloid: a nitrogen-containing organic compound of vegetable origin.

Alternate: where the leaves are inserted on the stem at different levels.

Anthocyanins: red or blue pigments found in flowers, fruits and leaves.

Anthracene (glycosides): compounds of plant origin composed of an aglycone derived from the anthracene nucleus linked to one or more sugars.

Auctor: the person who first scientifically described a plant species.

Berry: fleshy fruit containing small seeds or pips.

Bracteate: small leaf situated at the base of the flower or on its stem.

Calyx: lower part of flowers composed of sepals.

Capitule: compact agglomerate of flowers on the same receptacle.

Carpel: female reproductive organ of flowering plants.

Carbohydrates: compounds containing carbon, hydrogen and oxygen representing the most important of the energy-producing elements in plants. Sugars and starch are prime examples.

Carotenoids: yellow or orange-red pigments whose name comes from carotene which is the first name of the series isolated from carrot.

Decoction, Decoct: procedure which consists of boiling the plant in water for a certain time to give a decoction.

Dentate: very finely serrated or jagged.

Dioecious: plants with separate male and female flowers.

Drupe: fleshy fruit which surrounds a hard stone which contains the kernel.

Enzyme: an organic catalyst, protein in nature, which is produced by living organisms but which is capable of functioning outside of the cell.

Essential oil: complex mixture of odorous volatile compounds found in plants.

Flavanoids: yellow pigments composed of a sugar and an aglycone derived from a chromone.

GAP: Good Agricultural Practice.

Glomerulus: compact group of flowers.

Glycerides: principal constituents of vegetable or fixed oils composed of glycerol esterified by fatty acids.

Glycoside: a compound resulting from the linking of sugar and a non-sugar portion called an **aglycone** or a **genin**.

GMP: Good Manufacturing Practice.

Imparipinnate: leaf divided in several leaflets with an unequal number.

Infusion: a procedure which involves pouring boiling water over plants to obtain an infusion.

Population: Smallest functional biological unit where a change of genes takes place more or less regularly.

Proteins: nitrogenous substances of high molecular weight which on hydrolysis give amino-acids.

Proteolytic: enzymes (usually) which hydrolyse proteins.

Pubescent: covering of fine short hairs.

Raceme: group of flowers often loose and disordered.

Rhizome: perennial underground stem.
Samara: dry indehiscent, winged fruit which does not open, usually one-seeded, sometimes two-seeded.
Saponin: a glycoside composed of a sugar and an aglycone (genin) of steroid or triterpene origin.
Schizocarp: dry fruit in the *Apiaceae* family which splits at maturity into several indehiscscent one-cell carpels.
Sesquiterpene: compound formed from three isoprene units and a constituent of certain essential oils.
Sterol: an organic compound generally containing four carbon rings which is an aglycone in certain saponins.
Tannins: astringent compounds which have the ability to tan the skin and to combine with proteins.

B. Glossary of medical terms

Analgesic: pain-killer.
Ankylostome: parasitic worm found in the duodenum.
Anthelminthic: substance which causes the expulsion of intestinal worms.
Antiamoebic: substances which kill amoeba.
Antianaemic: substances preventing a reduction in haemoglobin levels in blood.
Antibacterial: substance which reduces the growth of bacteria and which can also destroy them.
Anticoagulant: substance which prevents the clotting of blood.
Antidiabetic: substance which reduces the level of glucose in blood (= hypoglycaemic).
Antidiarrhoeal: substance which stops diarrhoea - by reducing intestinal secretions - by reducing intestinal motility.
Antidote: a remedy to counteract the effects of a poison.
Antiemetic: an agent which counteracts emesis.
Antifermentation: substances which reduce the production of gas in the intestine.
Antiflatulents: substance which prevents gases being generated in the alimentary canal.
Antifungal: substance which reduces the growth of fungi and which can also destroy them.
Antiinflammatory: substance which reduces inflammation and the pain that results from it.
Antijaundice: substance which reduces the intensity of jaundice (icterus).
Antilithiasis: substance which prevents and destroys gallstones and kidney stones.
Antimalarial: substance which is used to fight fever due to malaria and which can eliminate the parasite.
Antimitotic: substance which prevents cell division.
Antinauseant: agent which counteracts the desire to vomit.
Antioedematous: substance which blocks or reduces the appearance of oedema.
Antiplatelet Aggregation: substance which prevents the aggregation of blood platelets.
Antipropulsive: agent which reduces gastrointestinal mobility.
Antipruritic: agent counteracting itching.
Antipyretic: substance which reduces fever.
Antirachitic: substance which prevents problems relating to the calcification of bone.
Antiseptic: substance which prevents the proliferation of pathogenic germs inside the body or on its surface.
Antisickling: substance which reduces crises due to sickle-cell anaemia.
Antispasmodic: substance which prevents spasms or cramps.
Antitrichomonas: substance which destroys trichomonas.
Antitussive: substance which prevents or reduces cough.
Anxiolytic: substance which reduces anxiety.
Arteritis: progressive destruction of the arteries.
Ascaris: parasitic worm found in the intestine (Roundworm).
Asthma: an allergy which results in respiratory difficulty.
ATC-Classification: Anatomical Therapeutic Chemical classification.
Blennorragie: gonococcal urethritis.
Bile Acids: acids liberated by the gall bladder which play a role in the digestion of fats.
Bilharzia: parasitic disease due to schistosomes which live in the human circulatory system.
Cholagogue: substance which increases the flow of bile from the gallbladder.
Choleretic: substance which increases the secretion of bile.
Cirrhosis: liver disease resulting from the progressive destruction of hepatic tissue.
Colibacillosis: inflammation of the intestines due to colibacilli.

Colitis: inflammation of the intestine.
Depurative: substance having the property of eliminating impurities from the body.
Diabetes: metabolic disease characterised by a raised blood sugar level.
Diaphoretic: having power to increase perspiration.
Diuretic: substance which increases the secretion of urine.
Dropsy = oedema
Dysentery: diarrhoea caused by a parasite or a bacterium.
Eliminative: to expel from the system.
Emollient: soothing and softening effect.
Expectorant: tending to facilitate expectoration.
Febrifuge: substance which reduces fever.
Galactogogue: substance which increases secretion of milk.
Gastritis: inflammation of the stomach.
Haemorrhoids: swelling of the veins of the anus and rectum.
Haemostatic: substance which stops a haemorrhage.
Hepatitis: liver disease caused by a virus.
Hepatoprotective: protector of essential liver functions.
Hypoglycaemic: substance which reduces blood sugar levels.
Hypolipaemic: substance which reduces blood lipid levels.
Hypotensive: substances which cause a fall in blood pressure.
Intraperitoneal: injection of a substance through the peritoneum.
Jaundice (icterus): yellow colouration of the mucous membranes and skin due to an increase in bilirubin.
Larynx: the modified upper part of the trachea.
Laxative: substance which accelerates intestinal transit and the evacuation of the bowels.
Leukaemia: blood disease characterised by a proliferation of white blood cells.
Malaria: parasitic condition due to the development and multiplication of hematozygotes of the genus *Plasmodium.*
Molluscicide: substance which has the ability to kill molluscs, e.g. snails.
Neuralgia: intermittent pain from the area of a nerve.
Oytocic: substance which causes contractions of the uterus and as a result hastens the onset of labour.
Otitis: acute inflammation of the middle ear.
Palpitations: rapid pulsation, especially abnormal.
Pancreatitis: inflammation of the pancreas.
Parasympatholytic: substance that antagonises acetylcholine.
Parasympathomimetic: substance with the same effects as acetylcholine.
Plasmodium: parasite responsible for malaria.
Prodromal stage: premonitory symptom.
Relaxant: substance which has the ability to make the body less firm, rigid or tense.
Rhinitis: inflammation of the nasal passages and respiratory tract.
Rubifacient: causing redness of the skin.
Sialogogue: an agent promoting the flow of saliva.
Sickle Cell Anaemia: hereditary disease due to the presence of a particular form of haemoglobin which causes deformities of the red blood cells.
Sympatholytic: substance which antagonises noradrenaline.
Sympathomimetic: substance with the same effect as noradrenaline.
Taenifuge: substance which is capable of killing and causing the expulsion of Taenia (tapeworms).
Ulcer: disease involving the chronic destruction of the lining of the stomach or duodenum.
Urethritis: inflammation of the urether arising from an infection.
Vaginitis: inflammation of the vagina.
Varices: abnormal swelling of veins.
Vasodilator: Vasoconstrictor: substances which have the property of dilating or constricting veins.
Vermifuge: substance which causes the expulsion of worms.

Index